Ali Cheshmehzangi
School of Architecture, Design
and Planning
The University of Queensland
Brisbane, QLD, Australia

ISSN 2731-6483 ISSN 2731-6491 (electronic)
Urban Sustainability
ISBN 978-981-96-6041-4 ISBN 978-981-96-6042-1 (eBook)
https://doi.org/10.1007/978-981-96-6042-1

This Springer imprint is published by the registered company Springer Nature Singapore Pte Ltd.
The registered company address is: 152 Beach Road, #21-01/04 Gateway East, Singapore 189721, Singapore

For Chongqing, the city that never sleeps, 11 December 2023 @3:30 a.m., Taken by the Author

To the City of Chongqing!

This goes to Chongqing and for the many times we met in and under different shades of blue. There will never be compunctions, but euphoric, eloquent, and vividly profound experiences. And I found this inconceivable arising new—a gem that rushed through many junctures with no pause, abeyance, and looking back. I got to know the familiar (and not similar) acquaintances there in 2005 (unconsciously) and later in 2015 (consciously) before I eventually got there in 2019 (in person)—and that connection is just extraordinary, beyond one can imagine.

So many years wasted before, and so many times we didn't see enough; or perhaps we did not seek enough. Some may have incredulity about all that happened but I still appeal to believe. Unity never neglects calamity, insecurity, and stubbornness; subsequently, healing is timely. An anonymous quote says "As the eagle finds its irie on the mountain peak, let your dreams find their foundation on the heights of your aspirations. May your irie be a sanctuary from which you launch into the vast expanse of your possibilities". After all, you have to meet your nemesis and face your ordeals—and that we did, just before reaching the redemption island. There we met again and again, just to know we are repeating the same cycle with all its old and new blunders, artifacts, and rituals that remain. Let's just say, we were there before we got here this round.

About This Book

Some books are written by mind, and some by heart. This book is written by mind and heart. City walks enable us to think and feel more intricately, and by living through them—hopefully less ostentatiously—through experiences and relations that genuinely matter. This sort of spatial syntagma is made of many correlations based on knowing what happens in cities, feeling what attributes are important, and articulating a set of relations to develop socio-spatial inter-relations and, ultimately, identities. In a way, "city walks" should empower cities to talk penetratingly while we hear, feel, and observe through experiences and thinking. Hence, city walks help the city talk.

The City of Chongqing means a lot to me. It is a city that allows you to fall in love with a sense of urbanity for city life and experiences. It is a city that creates a distinguishable set of relations, both transient and abiding in many ways. It is a city with a different tone of heartbeat during the day and the night, a sort of vibration that lives through many arteries at different nodes, settings, and connections. It is a city of multiple networks—different from many other cities—that never sleeps. Most importantly, Chongqing has a unique magnitude of everything that a city can offer. It is a city that can be experienced differently every time you visit it, and you can live through those diminutive experiences via greater connections and expounded senses that exist but have to be experienced first.

This is a book I have wanted to write for a long time. It maps Chongqing through many city walks and visits, and a new experience comes to life every time. The birth of such livelihood in mind and through various senses help us realise the fact that Chongqing is a sophisticated city of multi-layers, multi-identities, and multi-localities. Hence, the book is structured based on these three crucial aspects, including (urban) values, attributes, experiences, and livelihoods of Chongqing. This book is the first of its kind for a Chinese city—and what better city is out there as the start? Undoubtedly, many stakeholders in different parts of the world could learn from a city like Chongqing. Despite its vast size and large population as a metropolitan entity, in the form of a regional setting, Chongqing remains a pragmatic place with down-to-earth people, a prodigious food culture, and phenomenal socio-cultural values. With its prominent multiple layers, the city offers a range of identities and localities that are the backbone of Chongqing's image, experience(s), and life. This book aims to reveal

About the Author

Ali Cheshmehzangi is a Full Professor and Head of the School of Architecture, Design and Planning (ADP) at the University of Queensland. He has been in the World's top 2 per cent field leader since 2022, recognised by Stanford University. He is among the top 25 global scholars in the urban sustainability research area. Ali is internationally known for his scholarly contribution and extended work on climate resilience and sustainable urbanism research.

Ali has a wealth of experience, expertise, and a strong commitment to innovation and excellence in the fields of architecture, design, and planning, the three core disciplines of his School at UQ. With a career spanning over two decades, he has made significant contributions to the academic and professional communities, with a focus on sustainable and environmentally conscious design. Prior to joining UQ, Ali held several strategic leadership and senior managerial roles, such as Vice-President for International Engagement and Global Partnership, Founding Director and Head of the Center for Innovation in Education and Research, Head of the Department of Architecture and Built Environment, Founding Director of the Urban Innovation Lab, Director of a university-wide Teaching and Learning platform, Director of International Research Network for Rural and Urban Development, Head of Research Group for Sustainable Built Environment, co-director of university-wide research priority areas, Director of Center for Sustainable Energy Technologies, and Interim Director of Digital Design Lab. So far, Ali has published over 500 journal papers, articles, conference papers, book chapters, and reports. This book is his 39th academic book. Some of his previous books have received awards at the national, provincial, and municipal levels. He also has received international awards and recognition for his research on urban resilience studies and sustainability research, as well as a Vice-Chancellor's award for his impactful contribution to higher education.

Chapter 1
Chesh and Chongqing: Sun, Sin, Sanity, and the City that Sings

1.1 "Day One" in Chongqing

When I first visited Chongqing in October 2019, I felt an uncanny sense of familiarity—as if I had known the city long before setting foot in it. It was not a resemblance to any particular place I had been before, but rather a distinct feeling that was both unique and deeply perceptible. Let me refer to that moment as "day one"—a mild autumn day in October 2019, just months before the pandemic would begin to reshape our lives in ways we had yet to comprehend. As I wandered through the city, this notion of *placeness* became more pronounced. It was something that could only be fully grasped through an acute sense of place—an experience that felt alive, shaped by movement, interaction, and an immediate connection with the urban fabric. It was more than just the physical surroundings; it was a feeling of livelihood, an organic rhythm, and a set of relationships that formed almost instinctively between myself and the spaces around me. Chongqing revealed itself not just as a city, but as an evolving narrative—one that I was beginning to step into, observe, and, in some way, become a part of.

That *day one* was more than just the beginning of a journey—it was an entry point into an experience far beyond the city itself. It was about discovering the familiar in unfamiliar ways, everyday moments infused with a distinct character, a unique blend of flavours and rhythms that only Chongqing could offer. To truly grasp what this means, one must experience Chongqing firsthand. We had heard many stories—of its renowned cuisine, scorching summers, deep-rooted history, dramatic landscapes, diverse trades, and the warmth of its people. These stories were reassuring, yet they only hinted at what lay ahead. To *know* the city, we had to walk it—traversing its steep inclines and descents, crossing broad avenues and winding alleyways, meandering through bustling markets and tiny street-side shops, sharing breakfast with locals at a sidewalk stall. And all of this could unfold in just a few hours—immersing us in a city of multiple layers, identities, and localities. In that short time, we could observe the intricate interplay of people, food, commerce, and culture—the very elements

A. Cheshmehzangi, *City Walks in Chongqing*, Urban Sustainability,
https://doi.org/10.1007/978-981-96-6042-1_1

that shape Chongqing's essence. Yes, *people*—they are the heart of the city, the pulse that gives it life. And of course, *food*—it not only nourishes but defines the vibrancy and excitement of Chongqing. And *trade*—not just the transactions within modern shopping malls and commercial hubs, but the rich, textured economy woven into its streets, markets, and daily exchanges. Chongqing is more than a place—it is an experience, layered and alive, waiting to be understood through the simple act of being present.

On *day one*, you witness a city in contrast—where revitalized old towns pulse with new energy, while aging residential communities stand in quiet anticipation, waiting for the negotiations that will determine their fate. Here, the past and the future exist side by side: one resisting change, the other embracing it. Some residents hold onto their homes, tethered to a history woven into every wall and courtyard, while others contemplate the trade—a familiar, aging residence in exchange for a sleek, modern apartment in a towering high-rise. The transformation is neither abrupt nor absolute; it is a slow, unfolding dialogue between what has been and what will be. As the city stirs to life, you feel as if you are waking up *with* it—sharing in its rhythm, inhaling its first breath of the day. But in reality, the city never truly slept. Even in the quietest hours of the night, there was movement—night workers finishing their shifts, street vendors preparing for the morning rush, dimly lit alleyways where life continued in hushed tones. The presumed tranquillity was just that—a presumption. Beneath the surface, Chongqing remained alive, its pulse steady, its energy only momentarily subdued. This realization lingers as you step into the morning streets [11]. The city is not a place that simply begins and ends its day—it is continuous, layered, unfolding in cycles that do not conform to rigid timelines. And as you move through it, you begin to understand that experiencing Chongqing is not about observing change—it is about being immersed in its perpetual motion, its contrasts, and its unbroken narrative of past, present, and future.

Unlike its regional counterparts, Chongqing occupies a distinct position within China's urban landscape. It is one of the four core municipality regions, alongside Beijing, Tianjin, and Shanghai. However, unlike these other three, which are deeply entrenched in China's historical, political, or coastal economic fabric, Chongqing stands apart due to its unique geographical, economic, and administrative structure.

Being the sole core municipality in inland China, Chongqing has enough autonomy to run its urban and rural affairs under the direct supervision of the national government. The city has a strategic advantage due to its administrative status, which allows it to serve as a vital economic and logistical center for China's expansive western regions. The growth of Chongqing has been closely linked to internal migration, industrial transformation, and national policies intended to promote inland growth, in contrast to the coastal cities that have long profited from international trade [6]. Its urban form also makes this distinction clear. The highly organized and globally identifiable urban grids of Beijing and Shanghai contrast with the landscape of Chongqing, which defies accepted city planning principles. The city's topography, which is characterized by steep hills and the intersection of the Jialing and Yangtze rivers, forces a vertical urbanism that is uncommon elsewhere [4]. The city has an almost cinematic complexity as skyscrapers rise out of the cliffs, roads meander erratically through the rolling landscape, and entire neighbourhoods appear to float above or below one another. Chongqing's urban experience is multifaceted, layered, and ever-changing, so the idea of a flat, well-zoned city just does not apply here.

But it is precisely this intricacy that gives Chongqing its vibrancy. It is a city of flexibility rather than strict patterns. Its residents have mastered its complexities, riding monorails that cut through buildings, climbing staircases that serve as streets, and blending new and old areas into one cohesive urban fabric. This adaptability and resilience are indicative of a city that is constantly changing, striking a balance between its industrial past and its increasingly high-tech, service-oriented future. In many respects, Chongqing is a completely different type of city and not just another Chinese metropolis. A place where history and fast modernization coexist, where geography governs urban life, and where the extent of its transformation is as remarkable as the terrain it is situated on [10]. When you visit Chongqing, you will see a city that defies convention, celebrates diversity, and constantly reinvents itself in response to change.

1.2 Chesh and Chongqing

In this book, I, namely *Chesh*, will take you on a journey through Chongqing, the world's largest city by population. As an urbanist and urban designer, I have observed and experienced many cities across the globe, each offering a distinct sense of place. Walking through these cities, I seek to understand their spatial qualities and socio-spatial inter-relations, the environments they create for living and working, the vibrancy of their hubs, and how people move through and interact with their surroundings. My interests lie in the social dynamics of urban life, the presence of nature in cities, and the spaces that truly matter. I am fascinated by urban transformations, the transitions cities undergo, and the potential changes that shape them into better places to live, work, and thrive. Through this book, I invite you to explore Chongqing through my lens—its complexities, its contrasts, and its continuous evolution as a city unlike any other.

Chongqing is not like a typical city that one may have in min. It is a city that defies accepted urban logic because of its enormous size as well as its distinct topography, history, and quick changes. Chongqing's urban form defies conventional planning principles because it was built on steep hills and sculpted by two powerful rivers. Streets weave through layers of natural and constructed environments, frequently necessitating an experiential approach to navigation. They do more than just connect destinations. Chongqing's vertical urbanism creates a constantly changing landscape, with bridges, tunnels, staircases, and monorails all blending together in a complicated yet practical rhythm, in contrast to other cities where the skyline extends outward in predictable patterns.

I find the city's pulse—the way its residents move, engage, and define their spaces—to be more fascinating than its actual layout. In Chongqing, social life adjusts to the opportunities and limitations of the city, unlike in many other cities where social activities are governed by the architecture. People congregate in unexpected locations, such as rooftop terraces, riverbanks, and stairways, forming a dynamic urban fabric that is always changing in relation to time and purpose. I am especially interested in how cities like Chongqing change as they move through different stages of development and how these shifts affect how liveable they are. All of these characteristics are present in Chongqing, which is a city in motion and a living being moulded by its people, geography, and history. I encourage you to delve into its intricacies, contrasts, and innumerable layers through this book and learn what makes Chongqing unique among cities worldwide.

1.3 Chongqing: Sun, Sin, Sanity, and the City that Sings

Chongqing rises with the sun, but it never truly sleeps. It moves through foggy mornings where golden rays find it difficult to cut through the layers of a city made up of space, time, and memory [3]. Here, sky and concrete converse silently as tall skyscrapers vanish into the clouds and their reflections sway in the sluggish Jialing and Yangtze River currents. It is a city of paradoxes, where the past haunts Ciqikou's back streets while the future blazes ahead in neon-lit neighbourhoods teeming with aspirations. Against the monorails that wind through glass towers, ancient teahouses stand silently defiant, whispering the tales of those who have come, stayed, or gone. According to this paradox, Chongqing is neither wholly ancient nor wholly modern,rather, it is constantly evolving and never fully established.

Sin—the city knows it well. Not the kind found in scripture, but the indulgence of life in excess: the sweat of the worker ascending countless stairs to get home, the sound of mahjong tiles clattering late at night, the heat of a thousand chili peppers in a bubbling hotpot. Chongqing is a metropolis of appetite—a desire for sustenance, mobility, and survival. It does not ask for permission to be boisterous, wild, and to enjoy the mayhem it has created.

Yet, there is a strange sanity in this madness. Its disorder has a rhythm. Unspoken logic permeates the city, a system that functions not because it is meticulously planned but rather because it adjusts, flexes, and maintains equilibrium in the face of extremes. People here do not merely live in the city; they engage with it, reshaping their lives around its slopes, marketplaces, and quiet moments in between the whirlpools of activity.

And through it all, Chongqing sings. Street vendors' laughter, the sound of a ferry horn piercing the mist of a river, and the ceaseless clamour of life tumbling down its streets all contribute to its melodies. In the hands that create and destroy the city, in the bridges that connect not only land but also history and hope, it sings in the dialect that is spoken with ferocious pride. Sun, sin, sanity, and the city that sings its own tune—always changing but never changing—this is Chongqing.

1.4 A City of Heights, Depths, and the Spaces in Between

Chongqing is a city that does not unfold in a straight line. It defies the strict grids and predictable layouts of many contemporary metropolises by rising, falling, and bending around itself. Streets are stacked on top of one another in this vertical city, and entire neighbourhoods cling to hillsides as though in a precarious equilibrium between human ambition and the natural world. The topography of Chongqing is more than just a background; it actively shapes all facets of daily life, from how people move through the city to the layout of buildings and the development of communities.

The urban fabric of Chongqing has had to adjust to its undulating terrain, in contrast to many other Chinese cities where expansion has typically followed a horizontal sprawl. The city's infrastructure has had to evolve in unusual ways due to its deep valleys, steep inclines, and proximity to two important rivers, the Jialing and the Yangtze. Sometimes roads are stacked on top of each other in layers that do not make sense, and they twist and curve in unpredictable ways. In many places, staircases take the place of streets, transforming daily commutes into vertical excursions, and bridges span enormous distances between cliffs. In this context, elevation is a lived experience rather than merely a measurement of height.

The monorail system, which famously winds through residential buildings as if city planning and a child's imagination had blended together, is one of the most remarkable examples of Chongqing's urban ingenuity. The Liziba Station, where the train runs directly through an apartment building, has come to represent Chongqing's distinctive spatial solutions and serves as evidence that adaptation is not an exception but rather a requirement in this city. In order to make up for the difficulties of the terrain, the monorails, cable cars, and river ferries create a multifaceted transportation system that gives locals ways to move around a city where going from block to block can occasionally feel like climbing a mountain.

However, Chongqing's topography has an impact on its social and cultural life in addition to its physical infrastructure. Despite the city's enormous size, the steep hills and layered streets create unique microenvironments where neighbourhoods exist in a semi-isolated state, encouraging close-knit communities. In ways that defy traditional urban planning, local markets are frequently tucked away in tiny alleyways, stairwells serve as places for people to congregate, and rooftops double as courtyards. The idea of "in-between spaces" has a new meaning in Chongqing. The transitional spaces where life takes place—on the steps between two roads, on the terraces where the city appears to float over itself, and in the quiet areas amidst the busy chaos—are just as important as the actual spaces between buildings. Chongqing is a city that exists not only in space but also in layers of experience, memory, and motion because of the way height and depth, movement, and stillness interact to define its rhythm.

The distinctive topography of Chongqing shapes not only the city's skyline but also how visitors perceive it. In contrast to conventional urban grids, Chongqing develops vertically, with entire neighbourhoods clinging to steep hillsides, streets piled on top of one another, and staircases in place of roads. Because of this terrain, creative urban solutions have been developed, including multi-layered roads, ferries, cable cars, and bridges, as well as monorails that run through buildings. Chongqing's verticality affects social life in addition to infrastructure, creating closely knit communities in remote areas of the city. A rhythm of movement between heights and depths is created by markets, gathering places, and everyday routines that adjust to the physical landscape. This topography is more than just a problem; it is a characteristic that distinguishes Chongqing as a city of adaptation, resilience, and transition, where every step is not only forward but also up, down, and between layers of space, history, and human experience.

1.5 Rivers, Bridges, and the Eternal Flow of Time

For Chongqing, the Yangtze and Jialing Rivers are more than just geographical features; they are essential to the city's existence. From the city's beginnings as a vital trading post to its current position as a major economic force, these rivers have influenced every facet of the city. These waterways, which run through the center of Chongqing, both literally and figuratively define the rhythm of life. They represent time and the unceasing flow of history and are the silent witnesses to the city's numerous changes.

Chongqing's spine is formed by the Yangtze, Asia's longest river, which flows through the downtown area and links it to the outside world. In the past, its waters were teeming with boats that carried people, ideas, and goods, making it the entryway for trade, culture, and mobility. The Jialing River, which joins the Yangtze in the heart of the city, serves as a boundary and a complementary force, giving Chongqing's physical and cultural landscape more depth. For centuries, these rivers, which meandered through the rocky landscape of Chongqing, functioned as natural thoroughfares, transporting armies and traders, immigrants, and conquerors, all of whom left their mark on the city's identity. The rivers have always served as conduits for communication, from the earliest ferry crossings to the ports of today. The busy commercial ports that support international trade now coexist with the ferry docks where traders used to sell their goods [5]. Nevertheless, the rivers maintain a timeless quality in spite of their contemporary purpose. They talk about a long history that goes back through the dynasties, about revolutions that changed countries, and about the people who have travelled both real and imagined distances using these waterways. The rivers carry more than just water,they carry stories.

The most striking feature of Chongqing's relationship with its rivers, however, is its bridges. These majestic structures, which span the Jialing and Yangtze, are not only works of architecture but also representations of unity in a city that could be easily divided by geography. Because Chongqing is a meeting point of the old and the new, the local and the global, the past and the present, the bridges' intricacy and magnificence serve as metaphors for the city itself. The Chaotianmen Bridge, which links the city's central districts with the expansive tracts of land to the east, is one example of a spectacular bridge. Others are more modest, but they are nonetheless necessary for the city to function. They work together to create a network that unites Chongqing's disjointed districts and permits unrestricted movement of people, ideas, and business in addition to transportation.

The rivers, however, symbolize the unending passage of time and are more than just routes for travel. As they continuously erode and shape the land below, they serve as a reminder to the city of its dynamic nature. The restless, never-ending flow of the water reflects the ebb and flow of markets, the rise and fall of dynasties, and the profound currents of social and political change. The rivers of Chongqing, which carry the city's stories with them and preserve both its past and its future in their timeless embrace, endure as the world fast-forwards into the modern era. The Yangtze and Jialing Rivers are therefore more than just geographical features; they are the center of Chongqing's life, its pulse, its past, and its future. The bridges that span them are more than just means of transportation; they are vital connections that bind this enormous, sprawling city to its foundations, to the forces that have moulded it, and to the unending passage of time that continues to shape its development.

1.6 A City that Consumes and Creates

In many respects, Chongqing is a city of consumption—a constantly thriving engine that throbs with the passion of advancement and the hunger of its people. From the renowned Chongqing hotpot to the prodigious selection of street snacks that line its bustling streets, it eats a lot. In addition to using electricity to fuel its fast urbanization, the city uses energy to fuel its residents' ambition and drive. It takes up space as the skyline extends across the hills and upwards, and it takes up time as it moves forward continuously, its pulse echoing through all the layers of its streets, marketplaces, and neighbourhoods. However, just as much as it consumes, Chongqing is a place of creation. The city's identity is defined by this duality of creation and consumption, a complicated cycle that influences its past, present, and future.

In the past, Chongqing was a major industrial city that was created out of necessity during times of war and the early stages of China's fast modernization. Its economic engines are still shaped by its history as an industrial center. Chongqing's economy has historically been based on manufacturing, with heavy industries like steel, chemicals, and auto manufacturing playing important roles. The city produces a sizable portion of China's steel and electrical appliances and is home to some of the country's most significant factories, including those of the Chang'an Automobile Group. The city's development was made possible by this industrial base, which allowed it to grow and modernize rapidly.

However, Chongqing has changed along with China's economy. The city's industries have become more diverse, and its embrace of innovation and technology has made it a major player in the global digital economy [8]. Tech behemoths like IBM and Qualcomm, as well as a variety of start-ups, have been drawn to Chongqing's flourishing tech scene, which is supported by government incentives and investment in innovation [7]. Chongqing is rapidly emerging as a hub for technological innovation, from software development to robotics, and its work in automation and artificial intelligence is supporting the city's economic growth in the coming years. Chongqing's ability to adapt, innovate, and create is demonstrated by its transition from a heavily industrial city to one that strikes a balance between high-tech industries and its manufacturing heritage.

Yet, Chongqing's vibrancy is not only rooted in its economic engines but also in its cultural and creative expression [2]. Tradition meets innovation in the most surprising ways in this city, which finds beauty in the blending of the old and the new. The vitality of regional artists, craftspeople, and performers fills the city's streets, marketplaces, and public areas, preserving its cultural legacy while advancing it into new creative spheres. Every alley is alive with the buzz of creativity, every street offers a fresh chance to experience the arts, and every corner of Chongqing is a city that is continuously reinventing itself rather than just sitting back and watching.

This cultural dynamism is greatly influenced by the city's thriving street life. Chongqing is a city that thrives on the diversity of its expressions, from the bustling food markets where the aroma of spicy peppers and grilled skewers permeates the air to the neon-lit streets of downtown where fashion and modernity meet. One of the most unique facets of its identity is its food culture. From its fiery spices to its communal dining culture, Chongqing hotpot is more than just a meal; it is an immersive sensory experience that taps into every part of the city's identity. People from all walks of life congregate to eat, socialize, and create new experiences in the constant bustle created by street food vendors lining busy pedestrian streets and tucked away in alleyways.

At the same time, the city's expanding art scene demonstrates its dedication to creative industries. Chongqing provides a venue for modern artists to express themselves, from the new galleries in the art districts to the modernist sculptures that adorn public areas. Its cultural establishments, like the Chongqing Museum of Contemporary Art and the Chongqing China Three Gorges Museum, symbolize the city's reverence for its rich past as well as its dedication to artistic innovation in the future.

By embracing modernity and sustainability in its design, urban planning, and public space innovations, Chongqing also encourages creativity. The city's desire to blend modern architecture with an emphasis on public spaces and the environment is reflected in its constantly growing skyline, which is punctuated by dynamic structures like the Chongqing International Trade and Commerce Center and Raffles City Chongqing. Parks and riverbank promenades, among other green areas, provide equilibrium to the city's fast urbanization, demonstrating how Chongqing strikes a balance between its aspirations for development and environmental preservation.

By combining tradition and modernity, Chongqing not only grows but also constantly reinvents itself [1]. The city is continuously changing and evolving into something new thanks to the creativity and drive of its residents, who feed a cycle of creation and consumption. From the art studios to the factory floors, from the hotpot tables to the digital labs, Chongqing is a city that builds on the aspirations of its past to create not just objects but also a culture, an identity, and a future [9]. In this city, there is a chance for creation in every situation, every encounter, and every undertaking.

1.7 Unveiling Chongqing: The Purpose Behind This Journey

I asked myself this question repeatedly—whether flying over the city, sitting in a comforting pause by its riverside, getting lost in the bustling street life, or experiencing it time and time again. Not surprisingly, every time the answer was different in my mind! How come this city always manages to captivate me? Beyond its well-known hotpot and striking skyline, what else can be found in Chongqing? I frequently use frameworks of transformation, sustainability, and global trends to analyze cities in my work as an urbanist and urban designer. A different strategy, one that concentrated on the lived experience of the city, its pulse, and its people, was necessary in Chongqing, however, because of its raw energy and dynamic contradictions.

The aim of this book is to offer an intimate, firsthand exploration of Chongqing, a city that defies easy classification. This book is not about the theoretical foundations of urban transformation or about cities in general. This book is a single, narrowly focused case study, in contrast to my other published book that explore sustainability, the comparative analysis of city systems, or the wider effects of urban change. It transports you right into the center of Chongqing and captures the multi-layered reality of a city that is simultaneously traditional and futuristic, old and new. It is a personal investigation of how a city develops in front of you, using walking, observing, and reflecting rather than statistics or policies.

This book is based on my own city walks—those moments where you step off the map and let the city lead you. Through these walks, I aim to understand Chongqing beyond its surface-level features. To show you the city as I have seen it, I will lead you through its winding streets, over its rivers, and through its bustling marketplaces. Whether it is the historical significance of its rivers, the difficulties presented by its geography, or the rich cultural life that permeates every part of the city, each chapter explores a different facet of Chongqing's identity. Through these observations, we will also consider the wider ramifications of urban life, including how cities create and consume, develop and change, and incorporate the narratives of their inhabitants into their spaces.

This book stands out for its in-depth examination of a single city—not merely as a standalone case study, but as a dynamic, living being that is always changing, evolving, and redefining itself. Chongqing is the focus of urban studies, not merely the setting. By concentrating on Chongqing, I intend to shed light on the intricacies and paradoxes that come with living in an urban environment and provide insights into a city that is both ageless and undergoing significant change. The goal is to tell a story that is both intimate and educational, bringing readers into the city's facets from a lived, visceral experience rather than merely a scholarly one.

References

[1] TJ Campanella 2008 The concrete dragon: China's urban revolution and what it means for the world Princeton Architectural Press New York City
[2] Chabrowski I (2013). Representing and coping with early twentieth-century Chongqing: Guide songs as maps, memory cells, and means of creating cultural imagery. Cross-Currents: East Asian Hist. Cult. Rev 2(2):272–302
[3] Cheshmehzangi A (2021) Urban memory, experiences, and palimpsests: The continuity and discontinuity of places. Urban memory in city transitions: The significance of place in mind. 121–188
[4] A Cheshmehzangi A Dawodu A Sharifi 2021 Sustainable urbanism in China Routledge New York and London
[5] Cheshmehzangi A, Tang T (2022). Chengdu-Chongqing city cluster: From an economic zone to a major economic circle in the Western China. In China's city cluster development in the race to carbon neutrality. Singapore: Springer Nature Singapore, pp 77–92
[6] Cheshmehzangi A, Tang T (2024). Navigating China's urban revolution: From tradition to transformation. In China under construction. Springer, Singapore, pp. 1–10
[7] Rogers J (2008) A bull in China: Investing profitably in the world's greatest market. New York City: Random house trade paperbacks
[8] SM Walcott 2007 The dragon's tail: Utilizing Chengdu and Chongqing technology development zones to anchor west China economic advancement J. Chin. Econ. Bus. Stud. 5 2 131 145
[9] Wee HK (ed). (2021). The social imperative: Architecture and the city in China. Actar D, Inc.
[10] Zhang L, Zeng J, Dai L (2023, November). The scene construction of modern community: A case of the residential area of Wulidian Street in Chongqing, China. In: International Conference on urban climate, sustainability and urban design. Singapore: Springer Nature Singapore, pp 841–853
[11] Zhang X (2011). Carrying out modernity: Migration, work, and masculinity in China (Doctoral dissertation, University of Pittsburgh).

Chapter 2
Chongqing City Walks and Mapping the City

2.1 Walking Through the Layers of Chongqing

Chongqing is a city that reveals itself in layers, with new facets of its complexity being revealed at every turn. This point will be covered more extensively throughout the book. Chongqing unfolds in vertical sequences, requiring movement not only forward but also upward and downward, in contrast to cities with wide boulevards and predictable grids. Alleys twist and turn, streets wind erratically, and staircases appear where roads appear to end. Each one leads to a different perspective of the river, a bustling market, or an unexpected courtyard. Walking here is not a passive activity; rather, it is an interaction with a landscape that defies traditional urban forms. vibrant markets, and Chongqing's verticality from a pedestrian perspective.

Chongqing's topography defines how one experiences it on foot. The city is surrounded by water and perched on hills, making a simple walk impossible. A single stroll may seem like a journey across several urban eras coexisting within a few blocks, a descent into the city's older neighbourhoods, or an ascent into the sky. Navigating its layers evokes a feeling of discovery because every ascent or descent offers a new viewpoint, exposing details that are obscured from a distance [4] [6]. Walking in Chongqing is an immersive, fragmented, and frequently physically taxing experience that is not linear (See Figs. 2.1, 2.2, 2.3, 2.4, 2.5, 2.6, 2.7, and 2.8). However, it is also incredibly fulfilling, providing opportunities to interact with places and people that are hidden from those who only glance at the city.

On one walk, you may find yourself stepping off a lively commercial street into a quiet residential alleyway, where old men play chess under the shade of a banyan tree [1]. Smaller sounds of everyday life, such as conversations coming from open windows, the occasional clatter of cooking, or the rhythmic sweeping of a doorstep, take the place of the main road's hum (See Figs. 2.9, 2.10, and 2.11). It's a dramatic yet smooth contrast. The city creates a fluid, constantly changing urban experience by allowing its layers to blend into one another rather than being divided by sharp borders. Life here seems to be lived not only in the designated public areas but also

A. Cheshmehzangi, *City Walks in Chongqing*, Urban Sustainability,
https://doi.org/10.1007/978-981-96-6042-1_2

Fig. 2.1 The complex topography of the Chongqing is unique in many ways, through many bridges and the combination of old and new (*Source* From open source images supplied by Unsplash)

in the spaces between—on stairs, in secret courtyards, and along the winding streets that run like veins through the city.

Chongqing's street life is one of its defining characteristics. The city moves according to its own natural rhythms rather than rigid urban planning (See Figs. 2.12, 2.13, 2.14, 2.15, 2.16, 2.17, and 2.18). The rich scents of freshly steamed buns, sizzling skewers, and the distinct scent of Sichuan peppercorn fill the air as street vendors set up their makeshift stalls in the morning. Conversations take place everywhere, motorbikes and delivery carts are dodged by pedestrians, and markets spill onto sidewalks. In this city, the streets serve as places for daily routine, interaction, and exchange in addition to being routes for travel. Walking through them is a way to observe the city's character as it shows in the way its people behave, how its goods are arranged, and the unwritten laws that regulate its unofficial networks of community and commerce.

But as easily as one can become absorbed in the bustle of Chongqing's street life, one can also happen upon quiet times for introspection and solitude. In particular, the riverbanks provide a respite from the bustle of the city, where the constant flow of the water contrasts with the city's never-ending energy. Elderly locals congregate here early in the morning to practice tai chi, which mimics the rhythm of the Jialing and Yangtze Rivers with slow, deliberate movements. The mist that frequently hovers over the water causes fishermen's lines to vanish as they perch along the riverbanks. These private moments are long shadowed by the bridges above, which are enormous engineering achievements that serve as a reminder to tourists that human-scale experiences are still intricately woven throughout Chongqing's vastness (See Figs. 2.19, 2.20, 2.21 and 2.22).

Walking in Chongqing is, in many ways, an exercise in recalibrating one's sense of space and movement. The city defies preconceived notions about how an urban setting ought to operate and does not follow predictable trends. Sometimes a street

Figs. 2.2 Chongqing is known for its high density and high-rise modern zones that showcase the city's mixed urban landscapes (*Source* From open source images supplied by Unsplash)

seems to go somewhere, but it ends in a staircase that descends into a different level of the city. At other times, an alley that doesn't look particularly noteworthy at first turns out to be a bustling hub of activity, with food vendors, secret teahouses, or tiny shrines nestled within its nooks and crannies. Walking around here is so captivating because of this unpredictability. It is an investigation of how the city is lived and experienced on the ground, not merely a way to get from one place to another (See Figs. 2.23, 2.24, 2.25 and 2.26).

The idea that Chongqing is a city that was created in response to its surroundings rather than imposed upon them is further supported by the physicality of walking through the city. People have adapted to the challenges of the landscape over

Fig. 2.3 Chongqing offers a vibrant cityscape at night featuring a skyline of tall, illuminated skyscrapers with various logos and building types (*Source* From open source images supplied by Unsplash)

centuries, as evidenced by the steep inclines and layered neighbourhoods, making the city resilient and uniquely attuned to its setting [7]. Chongqing builds on itself, accumulating history, infrastructure, and everyday life in a manner that makes every stroll feel like an excavation of space and time, in contrast to cities that spread outward in planned grids.

There are countless opportunities for exploration in the city for both tourists and locals (See Figs. 2.27, 2.28 and 2.29). Whether it's a rooftop café with a view of the expansive skyline, an old bookstore tucked away under a bridge, or a row of houses

Fig. 2.4 The city is experienced through various verticalities and crossings over the rivers (*Source* From open source images supplied by Unsplash)

Fig. 2.5 The city's vertical urban landscapes in a hazy context, showing different urban land uses of commercial and residential (*Source* From open source images supplied by Unsplash)

perched precariously along a cliffside with balconies that seem to reach toward the river below, every stroll offers a fresh set of discoveries. To stroll through Chongqing is to experience a city that defies easy categorization. It is a city where every step unveils a new layer just waiting to be found, one that must be felt, experienced, and comprehended in motion (See Figs. 2.30 and 2.31).

Fig. 2.6 The city's vertical urban landscapes are astonishing features of Chongqing (*Source* From open source images supplied by Unsplash)

2.2 Mapping the Unseen: Beyond Conventional Cartographies

Although maps are supposed to depict a city, traditional cartography frequently fails in a place like Chongqing. It is impossible to accurately map the city's complexity onto a two-dimensional surface. In contrast to cities with radial layouts or orderly grids, Chongqing develops in layers that overlap, with streets stacked on top of each

Figs 2.7 The city benefits from multiple layers of mobility and transportation networks (*Source* From open source images supplied by Unsplash)

other, infrastructure woven through hillsides, and pathways that appear to defy urban logic. The experience of walking through this city—the weight of its verticality, the confusion of its winding alleys, the sense of being suspended between land and water, past and present, order and chaos—cannot be captured by a traditional map, which can only mark roads, bridges, and the path of the river.

One needs to look beyond linear routes and fixed coordinates in order to properly map Chongqing. It is a city of movement, sensations, and unspoken connections rather than merely points and lines [8]. On a traditional map, two locations may

Fig. 2.8 Chongqing's extended transportation networks include a monorail track that often runs through the buildings, which is a major feature of the city's verticality and cityscape (*Source* From open source images supplied by Unsplash)

appear to be just a few hundred meters apart, but in practice, getting there may require climbing innumerable stairs, taking a ferry, traversing a bridge, or navigating an alleyway that abruptly becomes an expressway. The experience of moving through Chongqing is not just spatial, it is also emotional, temporal, and intricately linked to how people engage with the city's infrastructure. Hence, our mapping approach here in the book is based on the actual experiences through what we refer to as "Chongqing City Walks" (See Figs. 2.32, 2.33, 2.34, 2.35, 2.36 and 2.37).

Figs. 2.9 The same views that could be seen vertically can also be seen horizontally through their spaces (*Source* From open source images supplied by Unsplash)

Fig. 2.10 The panoramic view of the city features numerous high-rise buildings in multiple parts of the Chongqing (*Source* From open source images supplied by Unsplash)

Fig. 2.11 The linkages between the two sides of rivers often help creating a metropolitan view of the city (*Source* From open source images supplied by Unsplash)

Figs 2.12 Like many populated metropolitans, Chongqing offers a range of large and expanded transportation networks and streets (*Source* From open source images supplied by Unsplash)

By converting landscapes into comprehensible diagrams, maps have been used for centuries to try and impose order on the world. This task is relatively simple in cities with well-defined grids. Chongqing, however, objects to these oversimplifications. Capturing a place that is constantly changing is a recurring challenge in historical maps of the city. The backbone of the city is formed by the Yangtze and Jialing Rivers, whose paths have changed over time, affecting urban settlements and trade routes. As new districts emerge, the footprints of old neighbourhoods have been either erased or

Fig. 2.13 Chongqing also offers a vibrant cityscape with its busy multi-level highway interchange and scenes of dynamic urban traffic flow (*Source* From open source images supplied by Unsplash)

repurposed. Infrastructure projects, such as tunnels buried into hillsides or monorails that wind through buildings, make it even more difficult to depict the city's layout in a single, static form.

Satellite imagery offers a bird's-eye perspective, but even this fails to capture the intricate interplay of life within the city's crevices. The experience of leaving a

Fig. 2.14 There is a sense of street and connection, and there is a sense of street life and connection, too (*Source* From open source images supplied by Unsplash)

Fig. 2.15 Perhaps once of the most famous street scenes in Chongqing is this view of yellow taxis lining up at the airport (*Source* From open source images supplied by Unsplash)

Figs 2.16 There are many places where verticality meets horizontality through streets and paths of the city (*Source* From open source images supplied by Unsplash)

station and finding yourself suspended high above the river on a platform that feels cut off from the street below is something that a metro map may depict as a neat and effective system of transit lines. Any attempt to depict pedestrian paths on a flat map is an exercise in oversimplification because they frequently exist at multiple elevations. It is possible for a road that seems to be continuous to be broken by a cliff and only connected by a secret flight of stairs. The topography of Chongqing

Fig. 2.17 The urban street scenes often come with steep stairways between medium to high-rise residential towers (*Source* From open source images supplied by Unsplash)

makes even GPS navigation difficult, frequently taking users to locations that exist in elevation but not in accessible reality. GPS navigation has completely changed our understanding of urban areas.

How else can the city be depicted if Chongqing cannot be adequately shown on conventional maps? One method is to map the flow of people, goods, and energy through the city rather than the fixed locations of streets and buildings. A foot traffic heat map would show unexpected pockets of activity in quiet, hidden staircases and riverbank promenades, as well as dense concentrations in markets and transit hubs.

Fig. 2.18 This unique view has become a major hub for places to see and experiences a real metropolis (*Source* From open source images supplied by Unsplash)

Fig. 2.19 Districts are defined and are characterised with a good mix of urban layers, including the past and the new (*Source* From open source images supplied by Unsplash)

By using sensory experiences rather than street names to direct pedestrians, a map of scent could follow the undetectable currents of smoke and spice that drift from food vendors. A sound map could depict a city that never truly sleeps by capturing the distant echoes of construction, the rhythmic chatter of vendors, and the hum of the monorail.

Figs 2.20 Places to live, hubs to visit, and destinations in the city (*Source* From open source images supplied by Unsplash)

Fig. 2.21 Chongqing's nightlife and street life is one of the unique characteristics of the city (*Source* From open source images supplied by Unsplash)

Fig. 2.22 Chongqing's nighttime cityscape features life, activities, and playful destinations across the city, showcasing a blend of modern and traditional architecture and unique visual experiences (*Source* From open source images supplied by Unsplash)

Figs. 2.23 Contrasting landmarks in an urban setting in Chongqing is a common visual experience (*Source* From open source images supplied by Unsplash)

Personal cartographies—hand-drawn maps based on lived experience—offer another way of understanding Chongqing. Despite their lack of technical accuracy, these maps capture something much more valuable: the individual's subjective perception of a location. Longtime locals may draw their daily itinerary, noting not only streets and landmarks but also the alley where a beloved noodle shop has been for decades, the hidden bench where one can sit and observe the river alone, and the corner where street musicians perform. The real spirit of the city, which

Fig. 2.24 Meanwhile, Chongqing offers a unique urban skyline at night with illuminated skyscrapers reflecting on a calm body of water, which enhanced the quality of relatively modern urban landscape offered by the city (*Source* From open source images supplied by Unsplash)

Fig. 2.25 Could this be the modern urbanity of the city? (*Source* From open source images supplied by Unsplash)

cannot be captured in kilometers, is revealed by these maps, which are brimming with memories and feelings.

2.2.1 Visual Representations of Chongqing's Complexity

Chongqing's historical maps offer insight into the city's changing character. According to ancient representations, the city's core is delineated by walls and gates

Fig. 2.26 There is life and urbanity on water and on many hills of the city (*Source* From open source images supplied by Unsplash)

Fig. 2.27 Street life is part of everyday life in Chongqing (*Source* From open source images supplied by Unsplash)

and is bounded by its rivers. Maps from the twentieth century show the rise of new roads, factories, and bridges—evidence of rapid industrialization [9]—as the city's borders became hazier as it grew. Modern digital maps, on the other hand, place a strong emphasis on infrastructure, highlighting skyscrapers and highways, but frequently fall short in capturing the natural development of informal settlements [5] and the hidden areas of daily life.

Artists and urban explorers have been trying to map Chongqing in unusual ways in recent years. In an effort to depict the city's verticality, some have produced

Figs. 2.28 Urban life is represented in various forms and in various places in Chongqing (*Source* From open source images supplied by Unsplash)

three-dimensional maps that stack neighbourhoods like sedimentary rock layers. Others have documented street scenes, doorways, and pathways that traditional maps miss using photography as a kind of cartography. These visual narratives provide an experience-based, as opposed to abstract, method of navigating the city. Digital simulations that simulate pedestrian movement across various elevations provide a particularly striking illustration of Chongqing's complexity. Users can follow the movement of people through the city with these interactive maps, which uncover surprising connections between seemingly unrelated areas. On a traditional map, a

Fig. 2.29 Urban life is also captured in everyday places between residential units, commercial hubs, and places of connections (*Source* From open source images supplied by Unsplash)

Fig. 2.30 In many pockets of the city, you could pass along communities that exist but remain hidden and undetected (*Source* From open source images supplied by Unsplash)

Fig. 2.31 More than diversity of culture, there is diversity in Chongqing's built environments (*Source* From open source images supplied by Unsplash)

Fig. 2.32 There is a sense of connectivity here through the relation between a particular use and street life (*Source* From open source images supplied by Unsplash)

Figs 2.33 A moment of pause is seen and experienced in various form in the city, where there is engagement in the public realms of the city (*Source* From open source images supplied by Unsplash)

Fig. 2.34 A moment of pause is also a playful moment where the scene conveys a sense of playfulness and everyday life next to each other (*Source* From open source images supplied by Unsplash)

Fig. 2.35 A moment of pause is also experienced on the street edge, where people gather, eat, talk, and enjoy an afternoon break (*Source* From open source images supplied by Unsplash)

Figs. 2.36 Living by the water and on the hills have strong sense of relations for the locals in Chongqing (*Source* From open source images supplied by Unsplash)

journey that looks broken up may actually be a continuous thread that is interwoven through underground walkways, bridges, and staircases.

Ultimately, mapping Chongqing means acknowledging that no one representation can adequately convey its spirit. The city is present in time, memory, and lived experience in addition to space [3]. Millions of people walk its streets every day, and cartographers are not the only ones who are continuously remapping the area.

Fig. 2.37 Despite its complexities and geographical features, Chongqing offers a unique urban life that is sensed throughout the community with pride and honour in everyday life (*Source* From open source images supplied by Unsplash)

Every action made here redraws the city in unique ways, producing maps that are imperceptible but intricately linked to daily rhythms.

Perhaps the most accurate way to map Chongqing is not through lines and coordinates, but through movement itself. The city is a dynamic map that needs to be experienced, sensed, and comprehended while moving. Each individual carries their own mental geography there, mapping their journeys based on their relationships with the surrounding spaces rather than on fixed points. In this sense, the real map of Chongqing is not one that can be printed out or affixed to a wall; rather, it is a map that is present in the minds of those who experience it, a continuously changing log of experiences, recollections, and future travels.

2.3 The City Pulse, the City Flows, and the City Life

Chongqing is not a city that simply exists—it moves, breathes, and pulses with a rhythm entirely its own [10]. Its transit systems are never quiet, its streets are never still, and its bridges are never empty. Here, the city is defined by the currents of life that flow through it rather than just by its structures or monuments. The distinctive topography of Chongqing, where streets wind around hills, bridges span the vast Yangtze and Jialing Rivers like arteries, and transit lines wind through the cityscape, occasionally gliding high above the water and other times piercing through buildings, creates the city's sheer energy. The constant flow of people, products, and energy that defines Chongqing is not merely a necessity,it is the city's defining feature (See Figs. 2.38, 2.39, 2.40 and 2.41).

To walk through Chongqing is to experience its layers. The streets don't follow the organized grids that characterize many contemporary cities, nor do they develop in a straight line. Rather, they wind and twist, sometimes disappearing into market alleys that seem to go on forever or ending suddenly in staircases that descend into secret courtyards. Every trip is an act of exploration for pedestrians, and every detour offers a surprising connection. A quick stroll could lead one past a peaceful temple courtyard, away from the bustling bustle of a packed food street, and onto a narrow pedestrian bridge where the only sounds are the wind and the distant horns of river ferries below. Each of these areas, which are characterized by movement and interaction, adds to the city's overall pulse (See Figs. 2.42, 2.43, 2.44, 2.45, 2.46 and 2.47).

In Chongqing, bridges are more than just steel and concrete constructions; they are vital links that keep the city united. Without them, the city's districts would be cut off from one another by the two great rivers, which would split it into disjointed islands. Some of these bridges offer stunning views of the city skyline as they soar above the water like sweeping, majestic engineering feats. Others, like the ancient

Fig. 2.38 Representing the old in the new form and space (*Source* From open source images supplied by Unsplash)

Fig. 2.39 Vintage shops try to preserve and demonstrate the recent past (*Source* From open source images supplied by Unsplash)

walkways that connect riverbanks and carry foot traffic across the water as they have for generations, continue to be modest. Crossing these bridges is a transitional experience, not only between geographical locations but also between various aspects of Chongqing's identity. A bridge could connect the city's historic center to its more recent growth areas, or it could connect a busy business district to a peaceful residential area. People's perceptions of their surroundings are shaped by these crossings, which support the notion that movement in Chongqing is a means of comprehending the city rather than merely a physical activity (See Figs. 2.48, 2.49, 2.50, 2.51, 2.52, 2.53 and 2.54).

Chongqing's transit systems, too, reflect its ever-moving spirit. For millions of people, the city's monorail system—a marvel of urban infrastructure—is an integral part of everyday life. In contrast to traditional underground subway systems, Chongqing's monorail travels above ground, tracing the natural contours of the area, winding around hills, and occasionally slicing right through the center of buildings. It provides a scenic and efficient urban experience, turning the commute into a voyage through the city's layers. The trains offer passengers a brief but close-up look at the life taking place below as they glide over busy roads, skirt the banks of rivers, and cut through crowded neighbourhoods. These high-altitude tracks have a poetic quality—a feeling of motion suspended in mid-air, neither anchored nor completely separated from the earth.

2.4 A City Captured: Visual Narratives and Urban Stories

One city that defies categorization is Chongqing. At least, that is how I have seen and experienced the city over many visits and explorations. In all those events, I tried to not just experience the city but to understand it fully. It sounds like a

Fig. 2.40 Celebrating the local lifestyles and culture is a main part of creating the nexus between the new and the old (*Source* From open source images supplied by Unsplash)

mission impossible, but that is how it has been every time I am back in Chongqing. I still believe the journey is yet to be complete. In reality, no picture, painting, or drawing can adequately capture its scope, vitality, and intricacy. Nevertheless, snippets of its essence—moments captured in time that allude to the innumerable layers beneath—are exposed through visual narratives. The city offers countless compositions of harmony and contrast, making it both a subject and a storyteller. Through its winding streets and broad boulevards, neon-lit skyline and foggy riverbanks, historic marketplaces and tall skyscrapers, Chongqing's story is told not through words but through pictures, captured gestures, and quick street-side glances.

Chongqing is a city that is constantly changing, as one can see when strolling through it. In this location, where a glittering glass skyscraper serves as the background for an elderly woman selling dumplings in a centuries-old alleyway, the

Fig. 2.41 Chongqing's food culture is [almost] second to none (*Source* From open source images supplied by Unsplash)

Figs. 2.42 Nature and water is almost the foundation of the city's image and urbanity (*Source* From open source images supplied by Unsplash)

Fig. 2.43 Nature and water also play a part in navigating through the city, having a better spatial syntagma of places to visit, see, and experience, and connect one side to the other in many parts of the city (*Source* From open source images supplied by Unsplash)

Fig. 2.44 Riverside walks and parks are part of the city's never-sleeping life (*Source* From open source images supplied by Unsplash)

past lingers in the shadows of the present. This juxtaposition—a fragile, weathered wooden balcony standing defiantly beneath the steel and glass of a recently built high-rise—might be captured in a single photograph. It is a city that builds on its history rather than erasing it, incorporating it into its current urban fabric like an incomplete painting that is always evolving. Every building narrates a tale of what was and is becoming, and every street corner contains a memory.

Photographs of Chongqing capture its rhythm as well as its physical form. Even the stillness of a photograph can convey movement, such as the slow ripples of a ferry moving along the Yangtze, the dance of steam rising from a hotpot restaurant, or the blur of a passing monorail cutting through a row of apartments. The energy of the city permeates even the quiet times and is not limited to the bustle of people on its streets. The city's reflective side and its breaks from the chaos are reflected in

Fig. 2.45 It is hard to miss the merge of hills/landscape with the buildings across many parts of the city (*Source* From open source images supplied by Unsplash)

Figs. 2.46 Nature is part of closed living environments as well as public open areas of the city (*Source* From open source images supplied by Unsplash)

Fig. 2.47 Nature is also a visible part of the city, alongside highways, in between tall buildings, and next to the rivers. This image captures an urban landscape with a mix of natural and human-made elements in one place (*Source* From open source images supplied by Unsplash)

Fig. 2.48 There is continuity of life in places where people meet, interact, and live (*Source* From open source images supplied by Unsplash)

a lone figure standing at the edge of a bridge, looking down at the endless waters below.

Urban sketches and photography are instruments of comprehension. They make visible what words frequently cannot: the closeness of everyday existence, the subtle urban movements that are missed until they are captured in a photograph or sketched on paper. An image of an old teahouse with men playing mahjong conveys a sense of custom, community, and generation-old rituals. A quick pencil drawing of a busy street market, bustling with people exchanging goods and greetings, captures the vibrancy that no statistic can measure. In this sense, visually recording the city turns

Fig. 2.49 An example of people's places in commercial podiums of residential towers above (*Source* From open source images supplied by Unsplash)

Fig. 2.50 Often we experiences people's places as places to eat and socialise across the city (*Source* From open source images supplied by Unsplash)

into a process of deciphering it and charting its unseen currents. I sometimes wish I could sit in many wonderful places in Chongqing for hours and days and capture the realities that belong to the city and its vibrant livelihood.

The real essence of Chongqing is found in its modest details, which go beyond the city's impressive skyline and the sight of bridges spanning enormous rivers. The pieces that make up the whole are the neon glow reflected on rain-soaked pavement, the peeling paint on a storefront, and the street vendor's patient wait while the city passes by. The goal of visual documentation of the city is to preserve the commonplace moments that collectively define urban life, not just to capture beauty

Fig. 2.51 A typical view of people in people's places that boost vibrancy of the city (*Source* From open source images supplied by Unsplash)

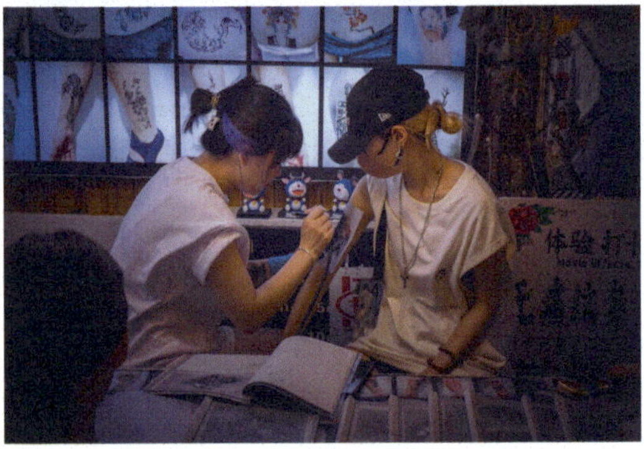

Figs 2.52 On the street, where life exists in different forms, temporary tattoos are part of everyday fun and joy of visitors to the city's vibrant street life and markets (*Source* From open source images supplied by Unsplash)

or spectacle. Chongqing, however, rebels against holding or having a single narrative. Because of its size, no two encounters with it are alike, and no two pictures convey the same meaning. One's perspective of the city is influenced by their position, whether they are looking up from a small alleyway in the old quarter or down from the observation deck of a skyscraper. Every viewpoint presents a unique Chongqing, or truth

Fig. 2.53 On the street, where life exists in different forms, there are many places to eat and gather, many places to have social life with friends and families, and many places where we could make memories that last (*Source* From open source images supplied by Unsplash)

Fig. 2.54 When the water goes down, new people's places are formed (*Source* From open source images supplied by Unsplash)

about who it is. A Chongqing of mystery and timeless beauty can be seen in this dawntime photograph of the riverbanks, where the mist whirls around the city's base. On the other hand, a nighttime view from a rooftop bar, with the skyline blazing with neon and movement, depicts a Chongqing of ambition and swift transformation.

Exploring the city is a visual storytelling exercise. In the continuous documentary that is urban life, every corner of a street offers a fresh composition, a fresh contrast, and a fresh frame. A photographer could document street vendors' morning routines, documenting how they set up their stalls, engage with onlookers, and move throughout the day. Drawing from a bench by the river, an artist might concentrate on the monorail's rise, following its lines as it moves through the cityscape like a living vein. By doing this, the photographer and the artist become more than just observers; they become active participants who influence how the city is perceived, comprehended, and remembered. When it comes to Chongqing, seeing and moving are inextricably linked. The essence of the city cannot be understood from a single perspective. The only way to start to comprehend how it is put together is to move through it—to walk its streets, cross its bridges, and descend its stairways. For this reason, traditional cartography cannot be used to map the city. The footsteps of people who walk through Chongqing, the memories created there, and the interactions that take place there are what truly map the city.

Perception and walking have a close relationship. A city's textures, sounds, and changing light are all experienced firsthand when one walks through it. It is to observe the city's changes in real time, such as the abrupt demolition of an old block, the appearance of a new plaza, and the seasonal variations that change the atmosphere of a well-known street. Additionally, a photographer maps the city with their photographs, an artist maps it with their sketches, and a walker maps it with their movements. Each is an attempt to frame a moment in a constantly moving, to capture something elusive.

Ultimately, a picture, a sketch, or even a book cannot adequately capture Chongqing as a city. It is too expansive, too multi-layered, and too flowing. But one can grasp its essence by capturing bits and pieces of it, such as the way light streams through a construction site's scaffolding, the way a child plays on an ancient temple's steps, or the way the river's surface reflects the city's nighttime glow. The visual narratives of Chongqing are dynamic; they are lived, experienced, and transformed daily. There is only one path forward for those who want to understand the city: to walk, observe, record, and participate in its developing narrative.

2.5 Discovering Chongqing: A Walk Through Time and Space

To walk through Chongqing is to enter a dialogue with time and space, to move through a city that is both immediate and historical, intimate yet overwhelming. Every action is not only a physical act but also an act of comprehension—an interaction with its networks, layers, and pulse. The city does not develop in a straight line; rather, it is revealed through its nooks and crannies, its streets that rise and fall, and its rivers that cut through the urban fabric. By doing this, Chongqing calls for experience as opposed to merely observation. To fully understand its depth, one must walk its streets, feel the weight of its past beneath their feet, see its unrelenting changes, and hear its voices in the echoes of busy marketplaces, quiet back streets, and bridges that span enormous bodies of water.

We can experience urban life in its most unadulterated and unrefined form by taking city walks. They lead us into the places where people live, work, and create meaning—beyond the glossy pictures of a city and its picture-perfect skyline. These walks are crucial to understanding Chongqing's complexity. The city moves, breathes, grows, and reinvents itself continuously; it doesn't just exist. Every walkway, stairway, bridge, and riverbank promenade has a unique story to tell, influenced by the people who use it and the forces of history that have shaped it. Chongqing is a city that is constantly changing, where high-tech innovation hubs coexist with historic teahouses and where modernity's unrelenting drive coexists with ancient customs.

The narrative of this book is organized around the very features that characterize Chongqing's urban experience, embracing the city's vibrancy. Through movement, observation, and interaction, it explores how a city is experienced, lived, and comprehended rather than just studying it. Through its spaces, flows, and contrasts, the book's structure reflects how one might actually experience Chongqing. It traverses the ways in which the city is viewed, charted, experienced, and eventually recalled. Each chapter unfolds like a journey through the city itself, from the act of walking through its complex layers to the invisible patterns that define its urban rhythms.

This book is also about urban identity—how a place shapes those who live in it and how, in return, people shape the place [2]. It draws attention to the fact that Chongqing is a city with many stories and realities that coexist. Even though the city has ambitious plans for infrastructure, it maintains the personal customs of everyday life. Although it is a city with a strong industrial sector, nature also permeates the urban landscape. It is a city where history is woven into the fabric of its streets rather than being restricted to museums. In order to portray a city that is both intensely personal and incredibly universal in its urban experience, I have attempted to capture these layers in this book.

This book provides academics, urban researchers, and city lovers with a fresh perspective on Chongqing, viewing it as a dynamic urban ecosystem rather than a static entity. It promotes a perspective on cities that embraces lived experience, sensory perception, and socio-spatial interactions, going beyond conventional urban studies. These city walks are an invitation to reflect critically on the built environment, to ask questions about how cities work and change, and to think about the human connections that make a city what it is. Chongqing is a city that defies easy definition because of its size, contradictions, and vitality. However, that's exactly what makes it interesting. You have to walk, feel, and experience this city. And by doing this, it shows itself not only as a location but also as a narrative, one that is being written with each step you take through its streets.

The book, as its sub-title entails, is structured in three parts, portraying Chongqing based on three evaluative narratives of (1) the multi-layered city, the (2) multi-identity city, and (3) the multi-locality city. All three are supported through the central approach of city walks. For Part 1, we delve into city walks that focus on four key characteristics and aspects/components of verticality, horizontality, streets and paths, and districts and zones. In Part 2, the focus on multiple identities through city walks and mapping Chongqing's urbanity, urban life, diversity, and relations. For Part 3, we explore in detail the key factors related to localities of Chongqing, such as old and new, food culture, nature and water, and people's places (See Fig. 2.55). The remaining chapters follow this structure before we conclude the book with a brief summary and a thorough reflection on a city that never sleeps.

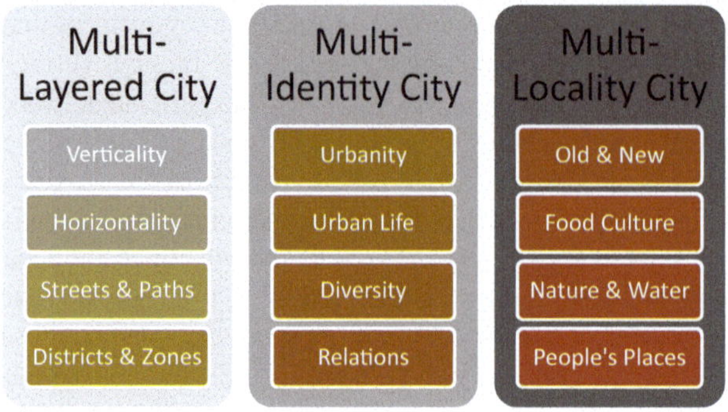

Fig. 2.55 Summary of the book structure

References

1 Chen J (2013) The politics of demolishing a neighborhood: Production and contestation of urban space in post-socialist China. State University of New York at Binghamton. Available online at: https://www.proquest.com/openview/0151d6e53d789563eb1ff85610015383/1?pq-origsite=gscholar&cbl=18750

2 Cheshmehzangi A (2020) Reflecting on urban identities. In: Identity of Cities and City of Identities. Singapore: Springer. pp. 133–178

3 Cheshmehzangi A (2023) Regenerating the local characteristics: the impacts on city life experience and identity of Shangri-La's old town. In: mapping urban regeneration: City life experiences in Yunnan, China. Singapore: Springer Nature Singapore, (pp. 219–237)

4 Cheshmehzangi A, Tang T (2024) Chongqing: Navigating rapid urbanization and regional development challenges. In: China under construction: Shaping cities through recent urban transformation. Singapore: Springer, pp 81–106

5 PC Huang 2011 Chongqing: Equitable development driven by a "third hand"? Modern China 37 6 569 622

6 Klingberg T (2014) Exploring place: how independent tourism is changing the politics of geographic knowledge in China (Doctoral dissertation, University of Colorado at Boulder)

7 Li Y, Cheshmehzangi A, He BJ (2023, November). A framework of urban renewal for leisure space in neighborhood streets: A case study of Shazheng street in Chongqing, China. In: International Conference on Urban Climate, Sustainability and Urban Design Singapore: Springer Nature Singapore, pp 693–705

8 Li Z (2015) The role of narrative in identity formation among new generation rural migrant women in Chongqing, China (Doctoral dissertation, Ohio University)

9 Shen Y, Sargani GR, Wang R, Jing Y (2024) Unveiling the spatio-temporal dynamics and driving mechanism of rural industrial integration development: a case of chengdu–chongqing economic circle, China. Agriculture, 14(6):884

10 T Yan J Shi 2015 Conservatism and evolution: educational continuation of Buddhist music in the Luohan temple of Chongqing city Can Soc Sci 11 1 104

Part I
The Multi-Layered City

Chapter 3
Verticality and the Metabolic City: Chongqing's Adaptive Urbanism in the Heights

3.1 Introduction: Walking the Vertical City

The cityscape of Chongqing is characterized by its striking topography (see Figs. 3.1 and 3.2), which has been moulded for centuries by the meeting of mountains, rivers, and steep slopes [4]. Because Chongqing extends upward and downward, it creates a spatial experience that is as much about vertical movement as it is about horizontal navigation, in contrast to cities that primarily expand outward. Exploring the city on foot reveals layers of culture, history, and urban adaptation, making it more than just a means of transportation. Walking through Chongqing is a unique experience that cannot be found in any other city. Urban walkers must constantly navigate stairwells, skywalks, twisting alleyways, and secret pathways that weave between high-rises and historic buildings, and the terrain necessitates a keen awareness of elevation changes.

The feeling of moving through several tiers of urban life is one of the most distinctive features of walking in Chongqing. One address may exist on multiple vertical planes in some districts; for example, what looks to be the ground floor from one street may actually be the tenth floor from another. The city's distinctive topography—roads are frequently stacked on top of one another, bridges connect at unexpected places, and entire neighbourhoods are built into hillsides—is the cause of this phenomenon. Chongqing's walkability is determined by the land's natural contours and centuries of adaptation to these difficulties rather than by traditional urban planning principles.

Chongqing's walking routes have traditionally followed the topographical lines of the city. The city's streets were mostly made up of riverbank trails, narrow walkways, and steep stairways prior to the current era of high-rise construction. This vertical complexity is evident, for instance, in the well-known Hongyadong district. Hongyadong, a multi-level commercial center where traditional architectural styles coexist with contemporary infrastructure, was once a collection of stilted wooden

A. Cheshmehzangi, *City Walks in Chongqing*, Urban Sustainability,
https://doi.org/10.1007/978-981-96-6042-1_3

Fig. 3.1 Chongqing's cityscape is characterized by its topography, high-rise buildings, and a sort of verticality as a mountainous city. It showcases a range of towers and landscapes. (*Source* Photo is taken by the Author in October 2019)

Fig. 3.2 New high-rise buildings often become the new part of the celebrated cityscape of Chongqing. (*Source* Photo is taken by the Author in September 2020)

houses along the Yangtze River. From the bustling marketplaces above to the waterfront views at the lower levels, walking through this district is like navigating a dense, layered space where different elevations offer distinct experiences.

Similar to this, Chongqing's urban layering is exemplified by Liziba, a neighbourhood well-known for its light rail that runs through a residential building. In this area, pedestrians come across a singular contrast between contemporary transit systems that blend in perfectly with daily life and historic staircases that lead to temples. Walking in this context involves more than just traversing space; it also entails simultaneously interacting with several urban narratives. Before escalators and cable cars made access easier, Liziba's steps were vital links for locals. Even now, they remain significant routes that provide a window into the past while blending in with modern transportation systems. City walks acquire even more historical significance for visitors to Chongqing's older neighbourhoods, like Ciqikou Ancient Town. Traditional tea houses, craftspeople, and tiny courtyards line the winding stone-paved streets, some of which date back more than a millennium. Although there is less verticality in these areas than in the city's contemporary center, the concepts of multi-layered movement are still present. Small staircases connect lower streets to higher terraces as alleyways wind through varying elevations. These historic streets maintain an organic, winding quality that reflects the city's natural landscape, in contrast to the organized grid systems of many contemporary cities.

Walking is still a major part of city life, even in Chongqing's most recent developments. The city's numerous elevated walkways and pedestrian bridges produce a special interaction between perspective and movement. There is not a single, continuous ground level as pedestrians move between shopping streets, transit hubs, and office buildings in Jiefangbei, the city's commercial center, thanks to a complex network of overpasses and underground passages. Instead of following a single path, walking here winds through a variety of heights, each of which offers a unique urban experience. Chongqing's public infrastructure emphasizes the connection between walking and the city's verticality. The city's many stairways, such as the well-known 18 Steps (Shibati), demonstrate how durable traditional pedestrian walkways are. Shibati, which was once a vital link between the upper city and the lower riverbank areas, illustrates how walking has long been the main way to get around the city's natural slopes. This ancient stairway still has value today, acting as a physical connection between Chongqing's past and present even as contemporary construction rises all around it.

Walking around Chongqing is a great way to practice spatial awareness and cultural immersion. Chongqing necessitates a different strategy than cities where walking is simple and linear. Unexpected connections between roads and buildings provide a sense of layered discovery, while seemingly short routes can be difficult due to steep ascents. Every stroll turns into an experience of the natural and constructed worlds, influenced by the city's constant play of depth and height. For me, walking through Chongqing is an act of interaction with its complex urban design. The experience of movement in this city is dynamic and multifaceted, whether one is traversing historic streets, touring contemporary skyscrapers, or descending staircases that have connected neighbourhoods for many generations. Discovering the various elevations, histories, and urban forms that contribute to Chongqing's identity is just as important as arriving at a destination.

3.2 Layered Urbanism: Mapping the Vertical Experience

Walking through Chongqing is like mapping a city that is made up of many layers of streets, bridges, stairways, and platforms that intricately intersect rather than being bounded by a single plane. Chongqing contradicts the idea that the ground level is a universal reference point, in contrast to cities. There are numerous grounds stacked on top of one another, each with its own unique atmosphere, experiences, and functions. We do not just walk through the city; we climb, descend, weave through underpasses, emerge onto unexpected terraces, and simultaneously find ourselves above and below the urban landscape.

Every step of a Chongqing city walk offers a different viewpoint on the built environment as it moves between these layers. In locations like Jiefangbei, we can be walking through a crowded pedestrian plaza with glass skyscrapers reflecting the midday sun one minute, and then we are descending into an underground commercial arcade that is crowded with commuters and vendors the next. As we emerge on the other side, we might ascend a flight of escalators that take us to a footbridge that is suspended over a highway, where we are abruptly treated to a higher view of the city. In the distance, the skyline rises further, with high-rises atop hills, their bases hidden by lower-lying neighbourhoods, and below us, the rooftops of older buildings unfold (Fig. 3.3).

The experience of walking in Chongqing is shaped not only by changes in elevation but also by shifts in urban rhythm. The streets in communities like Liziba are carved into the mountainside, requiring pedestrians to constantly adapt to the changing landscape. An unexpected grand staircase that plunges downward from a narrow alleyway between two apartment buildings may reveal a hidden market nestled between residential complexes. Intimate scenes of everyday life can be found in these urban pockets that are frequently hidden from view from the main roads above. Small food stalls operating at every level, laundry hanging from balconies at various heights, and senior citizens getting together for tea create a multi-layered sensory experience. The city's verticality is social as well as architectural, with various life strata developing on various levels (Fig. 3.4).

The way space is layered in some places, like Hongyadong, is almost theatrical. From its highest entrance, the riverbank-clinging complex looks like a traditional Chinese building, but as we move down its levels, it becomes a labyrinth of neon-lit hallways, eateries, and observation decks that float above the water. Each floor has a different soundscape; the upper levels are dominated by street noise, while the lower levels are filled with the river's hum and the muted voices of people strolling along the waterfront [7]. Walking through these areas allows one to experience Chongqing as a living, breathing structure carved into the natural contours of the land. The city is revealed gradually rather than all at once.

Fig. 3.3 For each new development, there are multiple layers added to the city, often including rooftops, roof gardens, multiple land-uses, etc. (*Source* Photo is taken by the Author in August 2024)

Another defining feature of navigating Chongqing is the way different modes of movement intersect. Although the city's extensive system of staircases, which date back several centuries, are still essential pedestrian routes, they coexist with contemporary mobility systems that improve the vertical experience. The contrast between the old and the new is particularly noticeable in locations such as the Huangguan Escalator, which is one of Asia's longest outdoor escalators. Locals still primarily use steep stairways to navigate the dense network of alleys that are accessible by foot from the escalator. When we exit the Yangtze River Cableway and return to the streets, we are instantly reintegrated into the pedestrian fabric, where footbridges and shortcuts through hillside corridors continue to dictate the rhythm of urban movement. In a similar vein, the cable car provides an aerial view of the city's layers.

Fig. 3.4 A typical example of residential buildings/complexes over a commercial podium. (*Source* Photo is taken by the Author in 2022)

Walking also reveals the way Chongqing's infrastructure negotiates its topography. Here, roads wind around hills, spiral upward, and occasionally seem to disappear into buildings, in contrast to grid-based cities where streets follow predictable patterns. In locations like the Eling neighbourhood, where a single address may be reachable from several levels based on which street we approach, this phenomenon is especially noticeable. Our sense of spatial orientation is continuously tested as we move through these spaces. When we look up, we may notice a road that runs parallel to the one we are walking on, but it is actually a continuation of the same road that loops around at a different elevation rather than a distinct street. Because of this spatial complexity, walking in Chongqing is never just about covering ground; rather, it is about comprehending how the city's various levels relate to one another and alter our perception of location (Fig. 3.5).

Fig. 3.5 An example of mixed verticality in central Chongqing with newly added commercial and mixed-use towers. (*Source* Photo is taken by the Author in August 2024)

Mapping the vertical experience through walking also means encountering moments of surprise. A peaceful staircase may abruptly lead to a panoramic overlook in some areas of the city, providing expansive views of the skyscrapers and rivers beyond. In other locations, what might seem like a dead end could actually be a secret passageway or unofficial path that leads to a completely different neighbourhood. These surprising findings support the notion that Chongqing is a city that must be experienced step-by-step, with each step exposing a new layer of its complex urban fabric.

The experience of strolling through Chongqing changes once more at night. Depending on our position, the city's elevation changes, and artificial lighting combine to produce unique visual experiences. Neon signage appears stacked against the hills from below, creating a luminous canyon of urban activity; from higher walkways, the lights' reflections dance across the river, highlighting the city's breadth and depth. Nighttime walking entails interacting with these disparate ambiances, as well-lit business districts give way to more subdued, dimly lit alleyways winding through older neighbourhoods. Chongqing's verticality is more than just a physical feature; it is a constantly shifting spectacle where time, light, and movement all affect how space is perceived.

For me, the city walks meant a lot in understanding the city's layered urbanism. More importantly, exploring Chongqing is a constant process of reinterpretation. The city does not appear in a clear-cut manner; rather, it develops gradually, requiring us to confront its intricacies. Every path we take gives us a fresh viewpoint, and every elevation change modifies how we perceive the areas we pass through. Walking in Chongqing is an immersive way to map the city's layered urbanism, whether we are traversing elevated bridges, underground markets, hillside streets, or historic staircases. We learn about the vertical nature of the city through this movement, not only as an architectural phenomenon but also as a lived experience that influences how people engage with their surroundings.

3.3 Everyday Metabolism: Life in the Heights

City walks of Chongqing resemble an invitation to experience life in the heights and at multiple layers, where the city is layered with human activity, culture, and business in addition to infrastructure. Moving through its vertical spaces is about more than just getting where you're going; it is also about experiencing the rhythms of daily life that vary in elevation. The streets of Chongqing are not just at ground level; they are on several levels, each of which displays unique social and economic dynamics. Walking turns into a means of interacting with these urban layers and charting the connections between the architecture of the city and the people who live there (Fig. 3.6).

Chongqing's elevated areas create a dense and dynamic urban metabolism, with homes, businesses, and markets stacked on top of one another. To walk through them is to see a city in motion. Street vendors erect their stalls along staircases that serve as pedestrian thoroughfares in the early morning as the first rays of light peek over the hills. Locals come down from their hillside homes to buy fresh vegetables and tofu from vendors positioned along the sloping path in locations like Shibati, the historic "Eighteen Steps". With freshly steamed baozi, sizzling jianbing, and bowls of hot rice porridge being served from makeshift street kitchens tucked into the stairways, the aroma of breakfast fills the air. Walking here is more than just a way to get around; it is a way of life, and is indeed, a way to take part in everyday activities and share space with merchants who have been selling their wares at the same elevation for generations.

The rhythms of life change as one moves upward. Office workers in Jiefangbei rush to their jobs in high-rises that appear to float above the streets on one of the many

Fig. 3.6 An example of life in heights in Chongqing with vast range of residential and commercial towers. (*Source* Photo is taken by the Author in August 2024)

sky bridges that cross the city. People move through the city on these raised walkways, which are suspended over-crowded streets and layered business areas, never coming into contact with what is typically thought of as the ground. A shopping mall's escalator leads to a pedestrian bridge, which connects to the mezzanine of another building, allowing for a smooth transition between levels. From here, one can gaze up at balconies with laundry hanging, a silent reminder of the residences tucked away within the city's vertical labyrinth, or down at street vendors selling roasted chestnuts below. All these are embedded within the city's vertical maze, flavoured with urban life that is mapped only through walking through the city and its fascinating urban environments.

As the day progresses, exploring Chongqing's heights provides an insight into the social and economic diversity of the city's residents [5, 8]. The contrast between movement and stillness is particularly noticeable in locations like Liziba, where a residential building is traversed by the well-known light rail. From the train station, commuters enter the narrow streets that meander through historic neighbourhoods where elderly residents play mahjong outside their houses [10]. Even though the city's infrastructure is modern, everyday life is still small and personal [2]. Walking here requires navigating a variety of areas, passing a modern café on one side and an outdoor noodle shop on the other where senior couples congregate while enjoying bowls of suanlafen, the area's specialty sour and spicy noodles. The distinctive urban character of Chongqing is derived from the interaction of the old and the new, the elevated and the ground-level.

Residential enclaves further up the hills provide tales of adaptability and resiliency. Homes in neighbourhoods like Eling or the upper reaches of Yuzhong District are constructed into steep slopes, and from a distance, their entrances may appear to be on the fourth or fifth floor. A narrow stairway may lead to a busy communal courtyard where neighbours converse, kids play, and fruit vendors sell their wares from carts positioned at the slope's edge. Walking here is a perspective exercise. These areas flourish in the nooks and crannies of Chongqing's verticality, ostensibly obscured from the city's main thoroughfares. Walking through them is a very intimate experience; around every corner, one discovers a new aspect of domestic life and a private moment between individuals who have adapted to the intricate geography of the city.

Even at night, the elevated spaces of Chongqing tell a story of layered existence. They narrate a tale of multi-layered life. Walkways that wind through various levels of activity are illuminated by Hongyadong's neon lights, which shine against the hillsides. While boats transport people across the Yangtze a few floors below, couples walk along riverbank terraces. There is an overlapping soundscape of urban energy as music wafts from bars positioned at various elevations. One can feel the city's vertical pulse as they stroll along these lit streets; the city is a metropolis that never truly settles into one dimension but instead alternates between heights, depths, and viewpoints.

The values ingrained in Chongqing's layered urbanism, in addition to its topography, are what give its vertical life its distinctive character. Here, walking is more than just a means of getting around or exercising; it is a means of comprehending the complex social structure that binds the city. It serves as a reminder that urban areas are interconnected through daily interactions, commerce, and movement. Generations of locals have built their lives in the heights of the staircases, skywalks, and secret passageways, which are more than just pieces of infrastructure. They are places of memory, continuity, and adaptation.

We encounter Chongqing's diversity through city walks, where verticality is a lived reality rather than merely an architectural element [6]. Through movement, the multi-layered interactions of its citizens, and the ongoing balancing act between ancient customs and contemporary changes, the city reveals itself. We get closer to comprehending not only the natural landscape but also the human narratives interwoven with it with each climb and descent, with each unexpected link between areas [1]. Exploring Chongqing's heights is a voyage through the many lives that contribute to the city's identity, not just a voyage through space [3]. To me, it has always been a way to explore places and see new views from various angles and levels of the city.

3.4 The Multi-Layered City: Then and Now

Walking through Chongqing is a journey across time, where each step through its multi-layered spaces reveals traces of the past embedded within the vertical urbanism of today (Figs. 3.7 and 3.8). The city's distinctive topography has long shaped its streets, stairways, and bridges into a complex system of circulation that defies traditional urban grids, dictating how people move through the city. The people of Chongqing traversed its steep terrain via stone-paved trails and narrow alleyways that connected riverside docks to hillside settlements long before high-rise buildings and monorails cut through the mountains. In addition to being routes of travel, these ancient routes served as conduits for cultural exchange as traders transported goods from the Jialing and Yangtze rivers to the busy marketplaces above. Relics of these old spatial structures can still be seen today as we move through the city, woven into the fabric of modern life but hidden from view.

An example of this is the historic town of Ciqikou, whose winding streets, dotted with structures from the Ming and Qing dynasties, still have the feel of a Chongqing prior to contemporary verticality. Stepping on worn stone slabs that have supported the weight of traders and travellers for centuries is a tactile experience that makes walking here an immersive act. The trail winds uphill past historic tea houses where senior citizens enjoy pu'er while taking in views of the river below. The district's multi-layered structure is still present despite the changes brought about by commercial tourism; wooden balconies loom over lower streets, and secret staircases connect them, providing unexpected shortcuts between levels. Navigating these areas is similar to how people have always moved through Chongqing—adjusting to the terrain, moving fluidly between elevations, and making it harder to tell the past from the present.

In addition to these historic pockets, the city's transformation into a contemporary metropolis has further solidified its reputation as a vertical city. As Chongqing's industrial core rapidly grew in the middle of the twentieth century, workers' housing began to rise in dense clusters up the hillsides, creating a unique pattern of terraced urban life. In older neighbourhoods such as Shibati, the layered remnants of this era are still visible: narrow alleys where laundry hangs between buildings, concrete staircases connecting communal courtyards, and small shops operating on several floors of the same building (Fig. 3.9). These areas, which coexist with the tall skyscrapers that now define the skyline, are still vibrant and essential to daily life despite the passage of time.

The city's vertical development is most noticeable in places like Liziba, where the renowned monorail runs straight through a residential building, producing an almost unreal contrast between habitation and infrastructure. There is a unique layering of urban functions when you walk beneath the tracks: commuters gliding past their

Fig. 3.7 An example of spaces in vertical urbanism characteristics of Chongqing, featuring multi-storey buildings with a rooftop garden and living spaces. (*Source* Photos are taken by the Author in different times, in September 2020 (left) and August 2024 (right), respectively)

windows, families living above, and pedestrians negotiating the steep sidewalks below. What distinguishes Chongqing is its multifaceted choreography. Different layers of life unfold simultaneously in this city, where movement is vertical rather than linear, and the way people navigate these levels on a daily basis shapes the lived experience of urbanism.

Walking around Chongqing still reveals strong continuities in how people interact with their surroundings, even in spite of these changes. Stairways and walkways still serve as common areas in older neighbourhoods like Eling and Huangjueping, where people naturally interact with one another while moving about. We might pass a mahjong game playing at the foot of a staircase during a morning stroll, with neighbours stopping in the middle of a step to greet one another. A hidden market with

Fig. 3.8 An example of spaces in vertical urbanism characteristics of Chongqing, featuring a similar pattern of spaces as shown in the previous figure, but in modern parts of the city, where rooftop gardens and spaces play a major part in creating vertical spaces in the city. (*Source* Photos are taken by the Author in different times, in September 2020 (left) and August 2024 (right), respectively)

its vendors set up along a sloping alleyway where customers weave between levels to peruse fresh produce may be reached via a side path. These encounters, which are made possible by the city's verticality, serve to further emphasize that Chongqing's urbanism is about the layered social and economic networks that flourish there as well as the physical elevation.

Even though the city has welcomed new construction and high-speed infrastructure, these conventional spatial structures still exist and have evolved rather than vanished. The logic of Chongqing's multi-layered movement has been expanded with the construction of sky bridges and elevated walkways, which provide new routes through the city without destroying the old [9]. The pedestrian bridges that

Fig. 3.9 Staircases in Chongqing are part of daily walking routes and connectivity. (*Source* Photo is taken by the Author in 2021)

span public squares, office towers, and shopping centers in areas like Jiefangbei serve as contemporary continuations of Chongqing's ancient streets—raised passageways that connect the city's various tiers without requiring pedestrians to return to a single ground level. Walking here seems to be an evolution rather than a replacement of the ancient circulation patterns, a modern interpretation of the long-standing transition between hillside and riverfront.

In this sense, Chongqing's vertical cityscape's past and present are not distinct but rather entwined, creating an urban story that keeps changing with every new construction. As new technologies and design approaches aim to close the gaps between Chongqing's historic foundations and contemporary skyline, the city walks may see even more integration between their various layers in the future. But regardless of how the city transforms, one truth remains: walking will always be the most intimate way to experience its depth, its complexity, and its human pulse. The multilayered character of the city is more than just an architectural feature; it is a way of life and a spatial reality that influences day-to-day interactions and the fundamental experience of moving through Chongqing.

As we conclude this exploration of verticality through city walks, it is evident that Chongqing cannot be comprehended from a distance or from a single point of view. Every step you take in this city reveals a new layer, a secret passage, or an alternative viewpoint on city life. The act of walking in Chongqing is an act of engagement with its past, present, and ongoing transformation, from sky bridges to ancient trails, from futuristic elevated corridors to staircases carved into hillsides. Numerous elevations, rather than a single horizon, define the city's identity, and each one provides a unique perspective on the complex, multifaceted terrain of this genuinely vertical metropolis.

References

[1] Chen J (2020) The legacy of the Chongqing cultural routes living heritage (Doctoral dissertation, Deakin University).
[2] A Cheshmehzangi 2014 Spatial syntagma and identity of a place: Sensing, relating to, and knowing a place J Hum Behav Soc Environ 24 7 799 810
[3] Cheshmehzangi A (2020) Reflecting on urban identities. In: Identity of cities and city of identities. Singapore: Springer pp 133–178
[4] Cheshmehzangi A, Tang T (2024). Chongqing: Navigating rapid urbanization and regional development challenges. In: China under construction: Shaping cities through recent urban transformation. Springer, Singapore (pp 81–106)
[5] Huang S (2018) Living in place: Rebuilding the relationship between people and place in Jiulongpo District, Chongqing, China (Doctoral dissertation). Available online from: https://digital.lib.washington.edu/researchworks/bitstream/handle/1773/42421/Huang_washington_0250O_18708.pdf?sequence=1
[6] Mirra F (2022). Urban imaginaries: Contemporary art and urban transformations in China since 2001 (Doctoral dissertation, Birmingham City University).
[7] Nanwani S (2024) Human connections: Teaching experiences in Chongqing, China and beyond. Suresh Nanwani. Printed in Singapore and The Philippines
[8] Roast AC (2019) Peripheral modernities: Urban imaginaries, housing, and informality on the edge of Chongqing (Doctoral dissertation, University of Leeds)
[9] Wang L (2023) The architecture of the collective: The block and ideas of the city in modern China. Available online from: https://dash.harvard.edu/entities/publication/d3ecb3d6-603a-4d1a-ba82-52c4e4e14003
[10] Yao C, Li G, Yan S (2024). Design strategies to improve metro transit station walking environments: Five stations in Chongqing, China. Buildings, 14(4):1025

Chapter 4
Horizontality Reimagined: Unfolding Spatial Continuities and Urban Expanses in Chongqing

4.1 Riverine Axes and the Horizontal Foundations of Chongqing

We get a better understanding of the Yangtze and Jialing Rivers' significant influence on the spatial dynamics of Chongqing as we walk along its riverbanks (See Fig. 4.1). These rivers serve as the city's horizontal foundations, determining historical settlements, economic lifelines, and cross-cultural interactions in addition to being geographical divides. The riverfront walks provide an open, spacious experience where the city unfolds laterally, in contrast to the vertical labyrinths of Chongqing's inner districts. Through the river's continuity, we can traverse Chongqing's multi-layered history, from the historic docks that once managed enormous volumes of people and goods to the contemporary waterfront promenades that encourage leisurely walks.

A particularly revealing walk takes us along the Nanbin Road promenade, stretching parallel to the Yangtze River, where one can experience the full breadth of Chongqing's urban contrast [4]. The city's longstanding status as a significant inland port is maintained by the enormous river that flows toward the horizon on one side, transporting cargo ships and ferryboats. A sleek, contemporary skyline rises above on the opposite side, contrasting modern urban spectacle with historic riverine commerce. Walking here is an exercise in comprehending the horizontality of Chongqing, where space is shaped not only by vertical ascent but also by movement along the water. Changing scales characterize the experience,one minute we are passing a peaceful teahouse with a view of the river, and the next we are dwarfed by the enormous network of bridges and highways that cross the water to connect far-flung areas of the city.

Exploring the historic docks and wharves of Chaotianmen, which served as the entrance to Chongqing's trade networks in the past, emphasizes the significance of these riverfront walks even more. We come across relics of the former port economy as we stroll through this area: stone steps that descend toward the water and were

A. Cheshmehzangi, *City Walks in Chongqing*, Urban Sustainability,
https://doi.org/10.1007/978-981-96-6042-1_4

Fig. 4.1 Horizontality of the older Chongqing as displaced on one of the walls in the city for people to see and appreciate the changes the city has gone through over the many years of development (*Source* Photo is taken by the Author from an image on the wall, in September 2020)

once used by workers transporting cargo from ships to markets above. These steps are now used as walking routes, where residents work out first thing in the morning and street vendors set up shop by the river to sell fresh fruit and snacks. Here, the past is still palpable—not as a static memory, but as an active, lived space where the function of the space is still defined by movement.

We take another important city walk along the Jialing River's banks, especially through the more sedate areas where older residential areas and traditional fishing communities still have a connection to the water. The banks of the Jialing provide a more personal, regional sense of horizontality than the opulence of the Yangtze-side promenades. A different rhythm of urban life can be observed when strolling through these neighbourhoods; narrow alleyways lead to courtyards that face the river, where senior citizens sit outside in the late afternoon to observe the passing boats. This is a distinct kind of Chongqing, where people's daily activities are still rooted to the same riverbanks that have formed the city for centuries, and horizontal space is about continuity rather than just connectivity.

Walking across Chongqing's many river bridges is another way to engage with its unique spatial layering. For example, from a high vantage point, the Dongshuimen Bridge provides a panoramic view of the city's two river systems, enabling us to visually trace the lateral spread of urban districts along the water's edge. Walking across these bridges offers a different experience than the constrained spaces of the inner streets of the city; the openness of the river below and the expansive sky above provide a sense of spatial relief that makes it possible to appreciate Chongqing's larger geographical structure [5, 11]. These instances highlight the city's horizontality, both in terms of its physical size and the way it encourages mobility through its riverine networks, which connect various urban experiences via walking.

Taking the ferry across the Yangtze is arguably one of the most immersive ways to experience the riverine horizontality of Chongqing. The slow motion of a ferry ride puts the traveller in a moment of transition between the city's layers, in contrast to

Fig. 4.2 An example of changing urban landscape in Chongqing (*Source* Photo is taken by the Author in October 2019)

the monorail or bridges, which highlight Chongqing's verticality. We see Chongqing as a city of flows as well as heights when we stand at the edge of the ferry and observe the river's endless expanse in both directions. Here, urban life is influenced by both vertical ascent into the hills and lateral movement along the rivers. Another walk begins as one reaches the opposite shore, unveiling yet another dimension of Chongqing's vast urban narrative.

We encounter Chongqing's horizontality as a crucial aspect of its urban identity as we stroll along the riverbanks, cross the bridges, and traverse the gaps between. The city is spread out along the riverine corridors that have shaped its growth for centuries, in addition to being built upward. Walking through these areas enables us to observe the invisible relationships that form the city, including the horizontal flows of people, products, and histories that continue to move through Chongqing's constantly changing urban landscape in addition to the tall skyscrapers that dominate its skyline (see Figs. 4.2, 4.5).

4.2 Bridges, Tunnels, and Transit: Weaving Chongqing's Horizontal Connectivity

No other form of transportation can compare to the immersive experience of walking through Chongqing, which reveals its multi-layered urban complexity. Although the city is undoubtedly vertical—many of the movement is defined by steep staircases, hillside elevators, and dizzying cable cars—there is another important aspect of

Fig. 4.3 Another example of changing urban landscape in Chongqing (*Source* Photo is taken by the Author in September 2020)

Fig. 4.4 Another example of changing urban landscape in Chongqing with residential buildings and connections alongside the compounds (*Source* Photo is taken by the Author in September 2020)

Chongqing to consider: its horizontal connectivity. We frequently move through the city without realizing it because of the extensive network of bridges, tunnels, pedestrian walkways, and transit corridors. This traversal of layered horizontality is

Fig. 4.5 Another example of cityscape with new residential towers and old-looking commercial and leisure units (*Source* Photo is taken by the Author in 2021)

not linear; rather, it is a process of navigating urban areas [6, 10] switching between different forms of transportation, and taking in the city at various speeds.

Crossing the bridges in Chongqing is one of the most remarkable ways to experience its horizontal connectivity. One such instance is the Dongshuimen Bridge, which spans the Yangtze River. When I first walked across it, I was in awe of the contrast between the openness of the sky above, the vibration of the steel beneath my feet, and the deep hum of cars speeding by. I can follow the river's course and observe how urban life develops along its banks from the bridge's unique panoramic view of the city, which contrasts with the congested streets below. From here, it is easy to see how Yuzhong District's high-rise clusters contrast with the more scattered neighbourhoods on the other side of the water. The walk itself is an act of observing the connections between the city's spatial layers as well as a way to cross.

Beyond the bridges, Chongqing's tunnels offer a distinct horizontal experience that is both concealed and vital to the city's mobility. I was walking through the city's older neighbourhoods when I came to a pedestrian tunnel under a main thoroughfare close to Jiefangbei. It initially appeared to be a straightforward underpass, but as I continued to walk, I came to understand that it was actually a whole microcosm of urban life. Inside, street vendors had set up shop, offering everything from mobile phone accessories to fresh fruit. As though it were an extension of their neighbourhood, elderly residents sat along the tunnel's walls and engaged in conversation. This was not merely a transitional space but a place where horizontal movement created its own form of urban interaction.

Navigation of Chongqing's extensive transit system, especially the monorail, which is both horizontal and vertical [9], is another aspect of walking through the city. Line 2, which passes through a residential building in the Liziba area, is one of the most well-known routes. I got off at the station the first time I rode this ride just to enjoy the experience on foot. As I stood outside, I saw the train leave the apartment building, glide smoothly over the street, and then bend in the direction of the next stop. This combination of living quarters and transportation is a prime example of Chongqing's multi-layered urban design. I walked under the monorail and followed the elevated tracks down the street, noting how businesses had adjusted to this new infrastructure—small food stalls set up in the shadow of the elevated tracks, cafes with balconies to watch the train. In a delicate urban choreography, the train's movement was reflected in the steady flow of pedestrians below.

I continued walking through Guanyinqiao, one of the busiest business areas in Chongqing [1]. Here, the complex system of pedestrian streets and raised walkways shapes the experience of horizontality in addition to the roads and transit. In contrast to bridges that cross rivers, these walkways link public plazas, office buildings, and shopping centers, enabling people to move between areas without ever coming into contact with the ground [8]. I experienced a feeling of fluidity as I moved through this network,one minute I was in a busy marketplace, and the next I was on my way to a peaceful rooftop park with a view of the city. Moving horizontally between levels provided a fresh viewpoint on how Chongqing develops not only upward but also outward, generating several levels of connectivity.

The streets of Ciqikou Ancient Town, which face the river, were the site of one of my most thought-provoking walks. This historic district creates a different kind of horizontality—one that is anchored in time rather than infrastructure—by following the land's natural contours, in contrast to Guanyinqiao's ultra-modern landscapes. The pedestrian walkways here resemble historical arteries as they meander through historic tea houses, traditional courtyards, and tiny lanes where craftspeople continue to work. As I moved through this area, I was reminded that Chongqing's horizontal connections are about continuity—using movement to connect the past to the present—rather than just efficiency and transit.

As I continued exploring, I reached Eling Park, another key space where Chongqing's layered urbanism is experienced on foot. The park itself requires an ascent due to its hilltop location, but once at the summit, the horizontal expanse of the city is visible. The city is simultaneously connected and fragmented as the bridges, riverbanks, and endless networks of roads and buildings extend outward. As I stood there, I reflected on all the places I had been—the pedestrian streets that unexpectedly connected neighbourhoods, the bridges that carried me across the river, and the tunnels that unveiled hidden urban life. Each of these encounters contributed to a larger picture, one that viewed Chongqing as a city characterized by its complex, multi-layered horizontality rather than merely being vertical (see Fig. 4.6).

When viewed through various horizontal movement modes, these city walks demonstrate how Chongqing is a city of transitions (Figs. 4.7 and 4.8). Walking is a way to discover the invisible relationships that influence urban life, not just a way to move through space. Chongqing's horizontality is a crucial component of its identity [3], whether it is demonstrated by the bridges that cross the wide waters, the tunnels that pierce the city's subterranean, the transit systems that hover over

Fig. 4.6 There is a visual connectivity between verticality and horizontality of the city (*Source* Photo is taken by the Author in October 2019)

the streets, or the pedestrian networks that wind through business districts. We can interact with the city on several levels by strolling through these areas, taking in not only its physical aspects but also the interactions and rhythms that give it life.

Fig. 4.7 The multiple of layers of a leisure and commercial street where I have walked up and down so many times. The street has changed so many times and is due to change again through a regeneration project. (*Source* Photo is taken by the Author in December 2024)

Fig. 4.8 Street edges and public art often create a sense of connectivity that is visible and perceptible through the horizontality of the city's main commercial and leisure areas. (*Source* Photo is taken by the Author in 2021)

4.3 Sprawling Plateaus and Peripheries: Walking the Expanding City

Walking through the outer edges of Chongqing, away from the dense urban core, we begin to experience a different spatial rhythm—one defined by sprawling plateaus, shifting peripheries, and the gradual transition from the tightly woven city center to the ever-expanding urban frontier. The outer districts develop in wider, more expansive ways than Yuzhong's condensed, multi-layered streets, where verticality rules every movement. Here, Chongqing's urbanism's horizontality is demonstrated by the way the city sprawls outward, taking up enormous tracts of land that were formerly rolling hills, riverine wetlands, or agricultural villages. Although these areas are frequently disregarded when discussing Chongqing's complexity, they are a crucial component of its urban narrative, which can only be fully understood by exploring its changing outskirts.

One such stroll started in the more recent construction of Liangjiang New Area, a sizable neighbourhood intended to serve as a major center for technology and the economy. Here, the urban fabric is expansive, gridded, and planned, in contrast to the historic core's natural chaos. I observed that the spatial dynamics were different as I strolled along its wide boulevards; the roads were wide, the sidewalks were ample, and the skyline, despite being dotted with contemporary high-rises, felt less confined. This is the new face of Chongqing, where people are given room to breathe rather than fighting for it. However, the walk uncovered surprising levels of continuity in spite of its modernity. Like in any of Chongqing's older neighbourhoods, I happened upon a group of elderly men playing chess under a shaded pavilion at a small intersection. I was reminded of the busy street corners in Jiefangbei by a vendor selling grilled skewers a short distance away. These commonplace elements, juxtaposed with a fresh urban landscape, demonstrated how urban growth involves more than just infrastructure; it also involves the relocation of accustomed lifestyle patterns.

Going farther out, I strolled for a while through Yubei District's outskirts, where the city gives way to semi-urbanized areas. The difference between the organic and the planned becomes more pronounced here. Rows of identical apartment complexes rise on one side of the street, their facades expressing the aspirations of a city that is expanding quickly. On the other side, there are still traces of previous communities, such as low-rise brick homes with laundry dangling from bamboo poles and tiny vegetable patches that obstinately cling to the edges of building sites. As I passed through this transitional area, I witnessed firsthand how urban growth creates a palimpsest of development by layering new realities on top of older ones rather than merely erasing the past.

One of the most remarkable walks through the growing outskirts of Chongqing was in Bishan, a neighbourhood that has quickly evolved from a traditional agricultural center to a major urban hub. I found this walk to be an intriguing contrast between the old and the new. I started in the town center, where walkability was encouraged by the design of a system of pedestrian-friendly streets. There were modern cafes, bookshops, and green areas lining the sidewalks; this deliberate urbanism encouraged slow, methodical walking. However, the urbanity started to fade away as I made my way toward the outskirts, giving way to open fields, isolated farmland, and a few ancestral homes. I came upon an old couple caring for a little vegetable garden in this liminal area, their small piece of land surrounded by tall buildings in the background. They told me that they had been in this neighbourhood for more than 50 years and had seen the city slowly but surely change the landscape as it approached them.

This experience, like many others along the city's outskirts, served as further evidence that walking is about more than just moving through physical space; it is also about seeing change as it happens. Walking around these outskirts, where industrial parks are built on what were once wetlands and new highways sever old farmlands, reveals the conflicts and compromises that characterize Chongqing's growth. In these outer districts, walking is about tracing new spatial orders—grids taking the place of organic alleys, commercial hubs arising where markets once stood, and pedestrian walkways planned rather than evolved—in contrast to the densely populated core, where mobility is frequently about navigating constraints.

As I walked through Jiangjin, another quickly growing neighbourhood on the western outskirts of Chongqing, I came across yet another level of urban growth. The horizontal scale of development was even more noticeable in this instance. The city seemed enormous, with streets that stretched well beyond direct view and open plazas that stood out from the crowded inner neighbourhoods. However, remnants of Chongqing's earlier spatial logic were still present here. I discovered a traditional food market in a tiny side street close to a recently constructed shopping center. The smell of fresh produce and the sounds of haggling vendors filled the narrow lanes. Even with urban planning's lofty goals, the unplanned and unplanned managed to survive.

These walks make it abundantly evident that Chongqing's horizontality is about the lived experience of growth rather than merely physical expansion (Fig. 4.9). The edges of the city are dynamic areas where urbanity is constantly being negotiated rather than fixed borders. As one strolls along these edges, one observes how the landscapes gradually change, how the old and the new coexist, and how resilient daily life is in the face of shifting spatial orders. New districts' broad boulevards and wide roads may imply an organized, forward-thinking urbanism, but the human aspects of this expansion are revealed by the in-between scenes, such as an old woman selling homemade dumplings by a highway overpass or a group of workers relaxing in the shade of an incomplete tower.

Fig. 4.9 Chongqing's horizontality is about the lived experience of growth rather than merely physical expansion. (*Source* Photo is taken by the Author in 2024)

Therefore, walking through the growing city is an act of engagement with both the present and the future of Chongqing. It enables us to observe not only the areas where the city has grown, but also the ways in which people have adapted, resisted, and rethought these areas. It demonstrates how Chongqing's character—its multi-layered, multi-textured, and intensely personal urban life—continues to influence and be influenced by the ways in which people navigate it, even in the face of these enormous new developments.

4.4 Public Spaces and Everyday Fluidity: Walking Chongqing's Social Horizontality

Walking through Chongqing's public spaces is an exercise in fluidity—of movement, of social interactions, of the spatial rhythms that define everyday life. This to me is unique in places like Chongqing. An experience that unfolds across plazas, riverbanks, and wide pedestrian avenues where the city stretches outward rather than upward is provided by the horizontality of public spaces, in contrast to the compressed verticality of the city's core, where space is layered in close proximity. As I explore these areas, I discover that Chongqing's horizontal urbanism is about more than just the physical spaces; it is also about how people live and move through them, transforming the vastness into something dynamic and unique.

Walking along Nanbin Road, a lengthy riverbank promenade that parallels the Yangtze River, was one of the most striking experiences that demonstrated this sense of social horizontality. The open space here is a stark contrast to Yuzhong's congested, cliff-hugging streets. I noticed how people moved through the area as I walked—young joggers pacing the river, families seated together on the broad stone benches that border the waterfront, and elderly couples taking leisurely strolls. Within the same horizontal plane, the space permitted a variety of social interactions and speeds. Walking in this context meant more than just getting around; it meant entering an urban environment that beckoned one to stay, to stop, to be a part of a constantly changing social landscape.

Further along, I entered Eling Park, one of Chongqing's largest and most historically significant green spaces. Walking in this park, which is perched on a narrow ridge, is important for reasons other than the expansive views of the city. It is the way people interact with the area—retiree groups tai chi in an open plaza, college students drawing in shady spots, and tiny tea vendors operating mobile carts. Every step you took through this area felt like you were experiencing the social rhythms of the city. Eling Park's pathways meander, encouraging impromptu detours, in contrast to the rigid, programmed spaces of contemporary urban design, where pathways dictate movement. As I strolled around the city, I noticed how the public areas permit a sort of daily improvisation, a balancing act between the planned and the lived.

This same feeling of fluidity permeated Chongqing's unofficial public areas, which serve as common areas even though they are not officially recognized as parks or plazas. Strolling through alleys and backstreets in older neighbourhoods like Shapingba and Huangjueping [7] exposed a distinct form of social horizontality. These were areas of change, where open courtyards became hubs of social activity, stairways became resting places, and streets became marketplaces. One evening, as I walked through the streets near Ciqikou, I stumbled upon a gathering of residents playing cards on a fold-out table placed right in the middle of the road. Traffic, both pedestrian and vehicular, just moved around them, adjusting fluidly, so there was no rush to clear the area. This flexibility characterizes Chongqing's horizontal urbanism, which is characterized by a shared negotiation of space where the lines separating the public and private spheres are constantly redrawn rather than strict spatial demarcations.

Perhaps nowhere is this negotiation more visible than in the lively public squares that punctuate the city. I was amazed by how space was used at Guanyinqiao Pedestrian Street, one of the busiest commercial areas. On the one hand, it was a place of hypermodern consumerism, with tall billboards and luxury brands. Fundamentally, however, the square served as an outdoor living space. Small crowds of people watched street performers, kids played by fountains, and senior citizens sat on benches and engaged in conversation. The area was adaptable, serving a variety of purposes throughout the day and changing its purpose based on the people using it. As I strolled around, I came to understand that Chongqing's public areas are dynamic settings that are shaped by how people use and inhabit them.

Another viewpoint of the city's horizontal expanses was offered by strolling through the riverbank neighbourhood of Chaotianmen, where the Yangtze and Jialing Rivers meet. A historic and modern gathering place, the expansive plaza extends toward the water. During the day, it is a bustling area with tourists admiring the scenery, snack vendors, and ferries arriving and leaving. But the place changes in the evening. A completely different atmosphere is created as groups congregate to dance in timed routines, young skateboarders seize the open space, and improvised food stands appear. Walking around here was like watching a performance of daily life, one that moves fluidly between various social registers.

It becomes clear from exploring Chongqing's public areas that the city's horizontality is a lived experience rather than just a result of architecture. It focuses on how people move through and change space, adjust to its openness, and form social bonds in the vast stretches of plazas, riverbanks, and unofficial meeting places. Public areas in horizontal urbanism encourage encounters, in contrast to vertical spaces where interactions are frequently divided within buildings. They turn into stages for life's routine performances, where prearranged uses give way to natural and impromptu interactions.

In many respects, Chongqing's public space's horizontality is a reflection of its citizens' adaptability. The city has maintained a shared space culture in spite of its rapid urbanization [3], where people constantly negotiate public life without formal planning restrictions. This is demonstrated by the coexistence of contemporary and traditional practices—upscale business districts that still allow for street vendors, planned parks that permit unofficial gatherings, and new construction that preserves traditional social interaction methods. Walking through these areas enables one to perceive the city as a living, changing thing that is influenced by its people rather than merely as a collection of planned settings.

The horizontality of Chongqing, as perceived from city walks, is evidence of the city's adaptability, fluidity, and intense sociality. The horizontal stretches of public space show a different side of the city, one that is open, participatory, and ever-changing, in contrast to the striking vertical ascents that define its center. We can map an urbanism that is defined by its people as much as its structures by strolling through these areas, which range from commercial squares to secret courtyards, from river promenades to impromptu street gatherings. For me, exploring Chongqing's social horizontality teaches us that a city is lived, not just constructed. The real essence of urban life is found in the open areas where people congregate, move, relax, and interact. These strolls through Chongqing's growing and changing public areas serve as a reminder that a city's horizontality is a social as well as a spatial reality—a dynamic web of exchanges, agreements, and shared experiences that shape the cadence of urban life.

References

[1] Bao HX, Li L, Lizieri C (2019). City profile: Chongqing (1997–2017). Cities, 94:161–171
[2] Cheshmehzangi A, Tang T (2024) 30 years of urban change in China's 10 core cities. Singapore: Springer
[3] Cheshmehzangi A, Willman E (2023) The city as palimpsest to redefine the role of architectural continuity in re-establishing urban identity in post-industrial landscapes. J. Urban Regen. Renew. 16(3):340–357
[4] Dan G (2018). Landscape design method of the urban waterfront slow-motility space (Doctoral dissertation, University of Pécs (Hungary))
[5] Han SS, Wang Y (2001). Chongqing. Cities. 18(2):115–125
[6] Jeffery KJ, Jovalekic A, Verriotis M, Hayman R (2013). Navigating in a three-dimensional world. Behavioral and Brain Sciences, 36(5):523–543
[7] Kobi M (2024) Urban energy landscape in practice: Architecture, infrastructure and the material culture of cooling in post-reform Chongqing, China. Urban Studies. 61(15):2979–2994
[8] Li Y, Cheshmehzangi A, He BJ (2023, November). A framework of urban renewal for leisure space in neighborhood streets: A case study of Shazheng street in Chongqing, China. In: International Conference on urban climate, sustainability and urban design. Singapore: Springer Nature Singapore, (pp 693–705)
[9] McNeill D, Connor A (2024). The logistical governance of vertical commuting in the central business district. Urban Studies, 00420980241263226
[10] Pasini R (2018). Landscape paradigms and post-urban spaces: A journey through the regions of landscape. Cham: Springer
[11] Xu Q (2017) The challenge of place identity in the making and experience of new Chinese urban square (Doctoral dissertation, University of Sheffield)

Chapter 5
Streets and Paths in Motion: Unfolding Chongqing's Layered Journeys

5.1 The Pulse of Chongqing: Mapping the Streets and Paths

Chongqing's streets and paths, though seemingly ordinary at first glance, pulse with the energy of the city itself [1]. These routes, which cut through and across the city's layered topography, are essential to comprehending how this vast metropolis moves, interacts, and changes. They are not just necessary pieces of infrastructure. We start to understand that every street in Chongqing, no matter how wide or narrow, crowded or peaceful, has a distinct story to tell—a microcosm of the city's social life and development.

The streets of Chongqing are not all the same. Adapting to the city's natural geography—characterized by its steep hills, meandering rivers, and vertical urban structure [7]—they twist, climb, descend, and entwine. It is evident when strolling around the city how the design of the streets and walkways represents this geographical diversity by providing a network of connected areas that support various forms of mobility. In one section, a broad avenue may accommodate the swift movement of automobiles and buses, but a short distance away, slender alleyways beckon pedestrians to travel more slowly and privately. As pedestrians, we frequently have to navigate these contrasts, acclimating to the frequent changes between the relatively quiet side streets and the busy rhythm of traffic (Fig. 5.1).

The first thing strikes me when walking through Chongqing is the varied pace and energy that each street exudes. With its buildings rising sharply on either side, their facades both modern and historic, you could find yourself swept along in the bustle of the city and the rush of motor traffic on one path. However, a short distance from these crowded streets, the landscape drastically shifts. It is easy to forget that you are in a city of millions when you are on the side streets. A different side of Chongqing is shown by these more sedate areas, where people congregate at tiny tea shops, where the sound of conversation blends with the lingering smell of street food, where the city's history seems almost real, and where the rhythms are slower (See Fig. 5.2).

A. Cheshmehzangi, *City Walks in Chongqing*, Urban Sustainability,
https://doi.org/10.1007/978-981-96-6042-1_5

Fig. 5.1 I have always remained critical about how street edges are poorly defined by south-north orientation of residential towers in many Chinese cities (*Source* Photo is taken by the Author in October 2019)

By taking these city walks, we begin mapping not only the actual streets but also the narratives that are woven throughout them. This is where the difference between the old and the new is most apparent. It is impossible to miss the gleaming new high-rises that dominate the skyline when strolling down the busy commercial streets in the downtown area. However, you will come across much older streets that have seen centuries of cultural interchange, trade, and change if you venture further into some of the city's more historic areas, such as the old neighbourhoods surrounding Jiefangbei. Despite being more dilapidated and less pristine, these streets are crucial to Chongqing's identity (Fig. 5.3). They serve as the foundation for the city's historical and cultural awareness.

In addition to serving as physical links between locations, these pathways serve as ties between the city's various communities and cultures. From the central business districts to the more residential or industrial zones on the outskirts, Chongqing's urban landscape is a patchwork of varied neighbourhoods, each with its own distinct identity. The streets provide a literal and symbolic route between the various tiers of the city, acting as the arteries that link these disparate locations. You can see how the city's past and present are entwined as you stroll through these streets. While a high-tech shopping center with gleaming glass surfaces reflecting the city's modernity emerges further down the same street, a traditional market with local vendors selling vegetables, spices, and street snacks may be encountered on a stroll down a street in the Yuzhong district.

It is clear from these city walks that the city's identity is dynamic [3, 4]. The public and private spheres, as well as tradition and modernity, are constantly interacting in the streets [2]. Observing the conflict between cars and pedestrians is one of the most

Fig. 5.2 The rhythm of moving from one space to another changes its speed as you walk through older areas and smaller courtyards of the city (*Source* Photo is taken by the Author in October 2019)

fascinating parts of strolling through Chongqing. In certain locations, it is evident that the streets were planned with automobiles in mind, but as we move through them, we observe how people have adjusted by making little areas on the sides of the roads, setting up makeshift marketplaces, or just finding places to stop and talk. The streets are not inflexible,rather, they change and evolve as people take charge of them, forming unofficial routes that represent the people's everyday lives in the city.

We also start to notice the complex connections between public and private areas when we walk. The streets of Chongqing, where public walkways meet private areas like residences, workplaces, and small stores, serve as a continual reminder of the city's multi-layered character (Figs. 5.4 and 5.5). This blending of private and public realms becomes especially apparent in residential areas where homes spill out into the street [6], with families cooking dinner in small outdoor kitchens or children

Fig. 5.3 The very busy commercial areas of old towns in Chongqing create a sense of place that takes you back to different timelines (*Source* Photo is taken by the Author in October 2019)

playing in the alleys. The real social fabric of the city is woven at these nexus points of private and public life. The streets of the city are transformed into places where people can interact in unexpected ways and not only get from one place to another.

We are reminded as we continue our strolls through Chongqing that the streets are not only useful thoroughfares but also actively contribute to the continuous development of the city. New developments are pushing the limits of what was previously familiar in some places, changing the landscape (Fig. 5.5). In others, where brick walls and old wooden doors evoke a bygone era, the streets maintain a feeling of continuity. However, the streets are always changing and evolving as the city adjusts to the shifting needs of its citizens, regardless of where we walk. This section concludes by highlighting how city walks offer a singular chance to observe how Chongqing's streets flow, not only in terms of movement but also in the way these areas transform into living narratives that convey the city's identity, culture, and history. Our walks help us to comprehend Chongqing's pulse, a rhythm that is constantly shifting but firmly anchored in the city's stratified landscape and its residents' relationship to the streets they traverse on a daily basis. These streets are ultimately more than just routes; they are the center of Chongqing, and every turn we take gives us a new perspective on the identity, history, present, and future of the city.

Fig. 5.4 A view of a Chongqing bridge under construction, embodying the city's constant evolution in linking its layered urban landscape. As streets weave through hills and rivers, pathways like this define the city's dynamic connections between past, present, and future. (*Source* Photo is taken by the Author in September 2020)

5.2 Hidden Paths and Unseen Routes: Unveiling Chongqing's Layers

Although we frequently find ourselves on the well-travelled streets and opulent boulevards that showcase Chongqing's modernity, the city has a completely different side that is concealed in the corners of less well-travelled streets. A completely different story that speaks to the core of the city's history, social dynamics, and ongoing change is told by these alleyways, side streets, and more subdued walkways. Even though Chongqing's skyline is dominated by its grand avenues and tall skyscrapers, it is frequently these lesser known but equally important streets that offer a closer look at the spirit of the city.

We are pulled into a world that seems timeless as we follow these undiscovered routes; one where the routines of everyday life happen more slowly, away from the bustle of traffic and the never-ending buzz of development. The fact that these side streets encourage you to stop and notice the little things that you might otherwise miss is one of the most fascinating things about strolling through them. The small, family-run stores nestled between apartment buildings, the worn murals on the walls, and the uneven cobblestones that have seen generations of foot traffic all provide windows into the city's soul and highlight the contrasts that characterize Chongqing's multi-layered urbanism. By exploring these undiscovered routes, one can learn about the informal uses of the areas and how the people of the city engage with their surroundings. The urban grid clearly defines the main streets, but the lesser-known

Fig. 5.5 In Chongqing, the river is not just a boundary but a guiding force, shaping the rhythms of urban life. This evolving waterfront, caught between construction and nature's quiet presence, reflects the city's perpetual dialogue between change and continuity. As new pathways emerge, they invite us to walk the edge of transformation—where history, water, and progress intertwine in an ever-flowing journey. (Fig. 5.5. 5.6) (*Source* Photo is taken by the Author in 2021)

streets run through the city like veins, guiding people through unofficial areas that straddle the boundaries between the public and private domains. These undiscovered paths are anything but anonymous; they are social hubs, where locals congregate to converse, where kids play, and where business frequently occurs outside of the official frameworks of marketplaces and malls.

The way that these smaller paths highlight the boundary between private and public life is among their most remarkable features (Fig. 5.7). The distinction between the two is frequently ambiguous in Chongqing. You might pass a store selling fresh vegetables down a small alley, only to see a family gathered around a small table,

Fig. 5.6 As day fades into night, Chongqing's streets hum with the quiet energy of movement and change. The old and new stand side by side—historic facades leaning into the glow of modern towers. This is a city in constant transition, where every step forward is a bridge between memory and possibility, between the familiar and the unknown. (*Source* Photo is taken by the Author in 2023)

eating, conversing, and living when the rear door of a residential building opens. These areas are in a transitional state, continuously changing and changing as locals adjust to the demands of the times. They are neither wholly public nor wholly private. The more casual, human side of the city—the shared memories and experiences that make Chongqing unique—is revealed in these areas.

Walking these untrod paths is especially fascinating because it highlights the differences between the old and the new. Life slows down and you are taken back in time to a previous stage of the city's growth as soon as you leave the bustling main roads. Narrow lanes wind between traditional homes in some of the older

Fig. 5.7 A street market nearby the Chongqing University is one of the city walk projects in the city to connect the public and visitors with localities of the city. (*Source* Photo is taken by the Author in December 2024)

neighbourhoods, which are characterized by low-rise structures and exposed wooden beams that pay homage to the city's rich past. These pathways bear the imprints of families who have resided here for many generations and are layered with memories of bygone eras. It is simple to picture the city before the steel and concrete of modernity changed its skyline. However, there is a feeling of change even in these older districts. Old buildings gradually give way to new types of development as new stores open and contemporary services are provided. Even these undiscovered routes are constantly changing, as the city itself is never static.

Exploring these lesser-known streets and alleys provides a distinctive perspective on the urban development of Chongqing. These calmer areas are not dominated by the demands of rapid development or the pressures of global capital, in contrast to the busy commercial areas. Rather than being the result of top-down policies or ambitious architectural plans, they represent the city's natural growth, where change arises at the grassroots level and is influenced by citizens' everyday activities. These areas demonstrate Chongqing's ability to strike a balance between its historical legacy and its contemporary goals, providing peaceful times of contemplation amidst the bustle of urban growth.

These undiscovered paths also highlight the various neighbourhoods that comprise the social fabric of the city. The streets in places like Shapingba's older neighbourhoods or the industrial outskirts have a distinct personality. Here, the paths frequently meet at local markets, small neighbourhood get-togethers, and outdoor areas that act as hubs for a variety of social events. A lively and intimate social atmosphere is produced by the sounds of conversation, children's laughter, and the continuous hum of street vendors. You get a sense of the city's strong sense of community—one that is based on interpersonal relationships, common experiences, and mutual support—by strolling along these paths.

Additionally, these more obscure, smaller routes act as links between different areas of the city. We can observe how these routes give us access to places that might otherwise seem cut off from the greater urban grid as we navigate Chongqing's neighbourhoods. Access to the more isolated areas of the city that are frequently missed in the haste of development is provided by certain paths that wind through terraced houses and lush greenery as they ascend the steep hillsides. For residents of these outlying neighbourhoods, these routes become vital because they link them to the center of the city and enable social and commercial interactions that are vital to the general operation of the city.

Furthermore, we can investigate Chongqing's evolution from an industrial center to a contemporary metropolis thanks to these undiscovered routes. In some neighbourhoods, we can follow the city's industrial history by strolling down former factory streets that have been converted into new commercial or residential areas. Although the architecture and layout of these areas still bear the scars of Chongqing's industrial past, there is a noticeable sense of reinvention—streets that once served the city's heavy industries now serve the needs of a growing middle class, and the way these spaces have evolved shows how adaptable the city is. For me, discovering these undiscovered routes reveals a city that is always changing. Despite differences in size and purpose, Chongqing's streets are all bound together by a common story of development, change, and adaptation. Where time slows down, where history and modernity coexist, and where the informal and formal collide in a dynamic tapestry of urban life, these lesser-known routes give us a glimpse into the city's many and frequently hidden layers. Exploring these undiscovered routes offers a more complex and personal view of Chongqing, one that is influenced by the daily rhythms of its residents and the ongoing blending of the old and the new.

The rich, frequently disregarded tales that serve as the foundation of Chongqing's urban identity are revealed to us through these city walks. These quiet, unobtrusive lanes might not catch the eye of the average person, but they provide a glimpse of the city's lifeblood in ways that no opulent boulevard or tall skyscraper could. They provide a deeper understanding of Chongqing's development and the social dynamics that contribute to its vibrancy and constant change, allowing us to see the city's soul.

5.3 Streets as Stories: Chongqing's Social and Cultural Pathways

We soon come to understand that Chongqing's streets are much more than just straightforward routes that get us from one location to another (Fig. 5.9). Stories abound on the city's streets, developed and evolved by centuries of social and cultural exchanges. In addition to reflecting the lives lived within the city's constantly shifting landscape, they serve as archives of collective memory and cultural practices. These city walks allow us to trace the layers of social and cultural significance woven into every street and walkway, in addition to simply taking in the physical infrastructure of the city.

We see firsthand how the streets in places like Jiefangbei and Hongyadong are much more than just places to get around; they are dynamic platforms where the city's daily routines and public rituals take place. As we stroll through these neighbourhoods, we observe how the streets serve as living manifestations of the residents' needs, wants, and cultural expressions. From the busy markets and colourful street festivals to the more sedate nooks where neighbours gather to chat or have a cup of tea, these areas are where Chongqing's social life takes place (Fig. 5.8). As living canvases, the streets each tell a unique tale about the people who use them, the events that have shaped them, and the changes they have experienced over time.

The feeling of cultural continuity that permeates the city's streets is among the most remarkable features of exploring Chongqing. We are encircled by layers of meaning and history as we pass through areas like Jiefangbei. In addition to being a contemporary urban center, this central business district also bears the remnants of the past. Old tea houses, tiny lanes, and small businesses that have withstood the wave of development are remnants of Chongqing's traditional culture, despite the area's streets being lined with shopping malls, tall office buildings, and flashy billboards. These older buildings serve as living examples of the city's social and economic development, and they are now dotted with contemporary architecture. Because the past and present coexist in a precarious balance, every step we take along these streets is a step through time.

However, streets serve as both sites of cultural reinvention and historical preservation. In Chongyadong, we observe how the streets actively participate in the ongoing cultural negotiating process of the city. By fusing historic architectural designs with contemporary business advancements, Hongyadong—a neighbourhood famous for its distinctive stilted buildings—has become a popular tourist destination. We can stroll through the old town and enter eateries, cafes, and stores that evoke the past while adjusting to modern preferences. Since traditional Chinese aesthetics are reinterpreted in ways that appeal to both locals and tourists from other countries, the very

Fig. 5.8 Street market in a more contemporary community creates a linear connectivity between multiple neighbourhoods here in the heart of Chongqing. (*Source* Photo is taken by the Author in August 2024)

layout of these streets speaks to the fluidity of culture. The continuous negotiation of cultural identity in a city that is changing quickly is reflected in these streets, where the interaction between the old and the new challenges our ideas of what it means to preserve cultural heritage.

As I walk through the city, the = streets also reflect the social tensions and divisions that permeate the city, even as they tell the tales of Chongqing's cultural development. We can sense the subtle variations in the ambiance, the types of businesses that line the streets, and the interactions we witness as we move through various neighbourhoods. While quieter, more residential neighbourhoods like the alleys in the Yuzhong District reflect a more local, working-class character, streets like Jiefangbei are heavily commercialized and draw middle-class shoppers and tourists. These divisions are social and cultural in addition to economic. From the wide boulevards of affluent neighbourhoods to the tiny streets where unofficial vendors set up shop to meet the needs of local communities, the routes we take expose the city's many levels of inequality.

It can be challenging to establish distinct boundaries between public and private areas in Chongqing because of the frequent intersections between them. Commercial activity clearly marks some areas, especially those in the city center, but other areas are much more fluid. For example, we observe how private life spills out into the streets when we stroll through residential neighbourhoods. Families congregating in public parks, kids playing on sidewalks, and locals conversing with neighbours on street corners are all common sightings. The distinction between individual and

Fig. 5.9 In every corner where streets bend into another urban hub, the city is well presented and defined. Here is an example of street edge and creating spaces that are not often on the map. (*Source* Photo is taken by the Author in August 2024)

collective life is continuously negotiated in this shared space created by these interactions, which blur the boundaries between public and private life. The streets themselves serve as social hubs where the daily rhythms of life are expressed, in addition to being physical links.

It is also evident from strolling through these streets that Chongqing's streets serve as platforms for the social drama of the city. We can observe the city's pulse as it changes from day to night in places like Jiefangbei. Early in the morning, the streets are crowded with people rushing to work, street musicians playing, and vendors selling hot breakfasts. The streets get busier, the crowd more varied, and the atmosphere more lively as the day goes on. As visitors and locals alike congregate to take in the neon-lit nightlife, meander along the river, and savour the regional cuisine,

Hongyadong's streets come alive with a distinct energy in the evening. These streets are dynamic, ever-changing to accommodate the needs and preferences of their users. We participate in the ebb and flow of everyday life as we move through these areas, becoming a part of this continuous social drama.

Stories from Chongqing's streets are not limited to the past; they also touch on the present and the future. We can feel how the streets themselves are changing as we continue to stroll around the city, how urban development, tourism, and globalization are reshaping them. In order to meet the needs of a society that is changing quickly, new neighbourhoods are popping up and the city's streets are being redesigned, even as the older, more traditional areas maintain their cultural significance. The way the city's social and cultural narratives are being rewritten is reflected in these changes to the street network, which go beyond simple aesthetic adjustments.

In many respects, Chongqing's streets serve as a canvas on which the social and cultural life of the city is painted. From the city's rich past to its vibrant present, every path we take reveals a new facet of its identity. The city's cultural life, interpersonal interactions, and stories are all told on its streets. We are experiencing the city's living history whether we are strolling through the bustling business districts, the more sedate residential areas, or the streets that are crowded with tourists. Through these city walks, we learn about the intricate relationships between the people who walk the streets and the stories that are woven into the very fabric of Chongqing.

In summary, Chongqing's streets function as social hubs and cultural markers. They are the sites where the public and private spheres converge, where history is preserved, and where contemporary life is negotiated. Strolling through these streets is an act of exploration that encourages us to consider the influences that have shaped the city and to interact with the dynamic, constantly shifting social dynamics that contribute to Chongqing's distinctive urban setting. We learn about the cultural and social narratives that make Chongqing a city of perpetual movement, change, and life by strolling through its streets, which tell the city's stories.

5.4 Streets as Vessels of Change: The Evolution and Future of Chongqing's Pathways

We are traveling through Chongqing's history and future as well as its physical landscape when we stroll through its streets. The city's streets are constantly changing as a result of new mobility tactics, changing social dynamics, and the pressures of fast urbanization. Strolling through these streets allows us to observe how Chongqing's urban fabric is changing, with each street reflecting the city's past as well as its future goals. This section will examine how Chongqing's streets, which have seen decades of growth, still act as catalysts for change, bearing the burden of the city's past as well as the hope for its future.

We are reminded that the only thing that is constant while strolling through the streets of Chongqing is change. The city's streets are constantly moving, from the narrow lanes of its historic districts to the wide, contemporary boulevards that cut through recently constructed neighbourhoods [9]. In constant change, the old and the new live side by side, sometimes in harmony, sometimes in conflict. We can observe directly how the city's streets have changed to meet the needs of growing urbanization as we stroll down streets like those in the busy downtown areas or the Yuzhong District. Older buildings are converted for contemporary purposes, and new buildings are constructed on the locations of old buildings. The streets have also changed,the once winding and narrow streets have been replaced by broad roads that are intended to move traffic quickly while preserving the traditional rhythms of pedestrian life.

The growing highway network and the integration of public transportation systems are two examples of how Chongqing's streets have been modernized. We observe that in addition to meeting the everyday needs of the locals, the streets now link Chongqing to the rest of the region and, more and more, the rest of the world. It is impossible to overlook the existence of contemporary infrastructure when strolling through neighbourhoods such as Jiefangbei. High-rise structures, retail centers, and office complexes that serve both domestic and foreign companies line the streets. The swift growth of these business areas involves more than just building new roads; it involves changing the city's fundamental identity. These streets are now hubs of international connectivity, attracting workers, investors, and tourists from around the world, making Chongqing a major force on the world scene.

Even so, Chongqing's historic streets continue to play a significant role in the city's identity despite its rapid development. We observe how these areas remain essential to local life as we meander through the city's more sedate back alleys. Even though they are no longer the primary conduits for the city's development, these streets are still vital to its social structure. The rhythms of daily life continue unhindered along these paths, where families congregate, and street vendors sell their wares. The spirit of Chongqing's history is on display in these less-travelled streets. The community centers, the older citizens sitting outside their houses, and the smaller, unofficial stores all contribute to a more personal, regional view of the city.

The way these streets serve as connectors—not just physically connecting various areas of the city but also tying communities together—is what makes them so intriguing. It is evident from strolling through Chongqing's streets that these avenues are being utilized more and more to cross social, economic, and geographic divides. Once feeling cut off from the city's more modern, commercial districts, the older, traditional neighbourhoods are now being reincorporated into the urban fabric. The blending of contemporary and traditional spaces in areas like Hongyadong fosters an inclusive environment where a variety of people—locals, visitors, and migrants—interact in common public areas. The streets themselves are evolving from places of separation to places of interaction, becoming symbols of connection.

The streets of Chongqing are changing to meet the demands of an increasing population as the city continues to grow. As we stroll through the city, we observe that green areas and pedestrian-friendly streets are becoming more and more important. A concentrated effort has been made in recent years to make the city more liveable by incorporating green corridors, improving public transportation systems, and making areas more pedestrian-friendly. Previously car-dominated streets are now being redesigned to accommodate pedestrians and cyclists. These modifications aim to create areas that encourage social interaction, support environmental sustainability, and enhance inhabitants' general quality of life in addition to enhancing mobility.

These streets will keep changing in the future to meet the city's expanding needs as well as its environmental issues. The way Chongqing's street system is evolving to meet the demands of urban resilience and climate change is among its most remarkable features [8]. We can observe the first indications of these changes during city walks along the Yangtze River and in the downtown area, from the installation of vertical gardens and green roofs to the redesign of streetscapes that place an emphasis on ecological sustainability. In order to make its streets and public areas more resilient to the effects of climate change, the city is not only looking outward but also inward. The streets of Chongqing are also playing a bigger role as venues for cross-cultural interaction. The city's streets continue to have their distinct local character while being influenced by global trends as it opens up to the outside world. In places like Jiefangbei, where traditional teahouses and street food vendors coexist with international retail brands, we can sense the meeting point of local customs and global consumer culture. Because of this dynamic, the streets become places for both cultural innovation and preservation, resulting in a rich, multi-layered cultural landscape. Today's pathways are not merely routes across space, they are routes through a dynamic city where communities, identities, and cultures come together and clash.

The city's continuous change will continue to influence Chongqing's street design in the future. These streets will bear the burden of the city's history while adjusting to the demands of a contemporary, international metropolis as it expands and transforms. Our city walks show us how these streets are evolving into places for social interaction, cultural exchange, and environmental sustainability rather than just being routes for physical travel.

For me, the streets of Chongqing are both products of the city's rapid growth and reflections of its ongoing evolution. We get a glimpse of the city's future as well as its past and present as we traverse these streets. Whether they are new or old, these streets are agents of change, bringing with them the histories of the city they help to shape and the people who use them. The intersection of mobility, social interaction, and cultural exchange will continue to shape Chongqing's streets in the future [5]. We are actively taking part in the city's transformation through our walks, not just witnessing it. As the city moves forward into an ever-changing future, Chongqing's streets will remain places of movement, connection, and change.

References

1. Björner E (2017). Imagineering place: The branding of five Chinese mega-cities (Doctoral dissertation, Stockholm Business School, Stockholm University)

2. Calhoun C (1998) Community without propinquity revisited: Communications technology and the transformation of the urban public sphere. Sociol Inq 68(3):373–397
3. Cheshmehzangi A (2020) Reflecting on urban identities. In: Identity of cities and city of identities, Singapore: Springer. pp. 133–178
4. Cheshmehzangi A (2023) Regenerating the local characteristics: the impacts on city life experience and identity of Shangri-La's old town. In: mapping urban regeneration: City life experiences in Yunnan, China. Singapore: Springer Nature Singapore pp 219–237
5. Cheshmehzangi A, Tang T (2024). Chongqing: Navigating rapid urbanization and regional development challenges. In: China under construction: Shaping cities through recent urban transformation. Springer, Singapore. pp 81–106
6. Jiang M, Nakajima N (2023) Chongqing People's Square after 1997: Situated publicness of municipal squares in reform-era China. Urban Research & Practice 16(4):489–517
7. Liu Y, Yue W, Fan P, Zhang Z, Huang J (2017) Assessing the urban environmental quality of mountainous cities: A case study in Chongqing, China. Ecol Ind 81:132–145
8. Lu H, Zhang C, Jiao L, Wei Y, Zhang Y (2022) Analysis on the spatial-temporal evolution of urban agglomeration resilience: A case study in Chengdu-Chongqing Urban Agglomeration, China. Int. J. Disaster Risk Reduct 79:103167
9. Xu Q (2017) The challenge of place identity in the making and experience of new Chinese urban square (Doctoral dissertation, University of Sheffield).

Chapter 6
Districts and Zones Unfolded: Walking the Layers of Chongqing

6.1 Mapping the Chongqing's Fabric: Distinctive Zones and Their Boundaries

The vast urban landscape of Chongqing presents a complex patchwork of zones and districts [9], each distinct yet connected, creating the city's distinctive fabric. As soon as I started these city walks, it became evident to me that every district has a unique rhythm, culture, and character that are influenced by the local population and geography. As I moved through these areas, I came to understand that their borders are not just physical, they are dynamic, constantly changing, and impacted by social, economic, and historical factors. As I continued to walk, I became more aware of how these lines are not always distinct; they frequently overlap, change, and occasionally even run counter to one another.

Consider the trip through the heart of the city, which begins at Jiefangbei (Fig. 6.1). I was struck by how this neighbourhood embodies the core of Chongqing's business and economic activity as I made my way through its busy streets. The streets are lined with upscale shopping centers, neon signs, and tall skyscrapers that depict modernity and global connectivity. But as I kept walking, I noticed that I was moving into the district's older areas, which had local markets, tiny alleyways, and the aroma of street food that suggested a bygone era. These two worlds appeared to coexist in a harmonious tension despite their stark contrast. The line separating the old and the new was not as sharp as I had thought. Rather, it had the feel of a living, breathing thing that was always changing and evolving. The district boundaries of Chongqing are essentially as follows. The rhythms of everyday life constantly negotiate them; they are not fixed. I now have a fresh perspective on how urban areas are mapped, not only by city planners but also by the people who live, work, and travel through them, thanks to my experience exploring these areas (See Figs. 6.2 and 6.3). The city's layout has an organic feel to it, a developing process that becomes apparent only when one interacts directly with the streets. Districts and zones in Chongqing

A. Cheshmehzangi, *City Walks in Chongqing*, Urban Sustainability,
https://doi.org/10.1007/978-981-96-6042-1_6

become more than just geographic divisions; they serve as archives of the social dynamics, history, and culture of the city.

The lines separating the residential and commercial areas became more obvious as I left Jiefangbei and headed toward Yuzhong, but even in these areas, the changes were

Fig. 6.1 Jiefangbei's cityscape is a visible part of the city's urban character. (*Source* Photo is taken by the Author in October 2019)

Fig. 6.2 The river and bridge connections create a sort of defined arrangement for the city's districts and zones. (*Source* Photo is taken by the Author in October 2019)

Fig. 6.3 Here, there is a sense of divide and connectivity at the same time. You can see where people may work and live, while you are aware of differences across the city. (*Source* Photo is taken by the Author in 2021)

fluid. Narrower streets replaced the wide boulevards, and the bustle of pedestrians muffled the hum of traffic. We walked through areas where more recent, contemporary buildings coexisted with older residential buildings. The streets seemed more personal and less hurried. Here, I could sense the bustle of daily life: elders conversing on street corners, kids playing in tiny alleyways, and families preparing meals in their kitchens. Far from the commercial glamor of the central districts, these areas provided a window into the city's human scale (See Fig. 6.4, 6.5 and 6.6)

What fascinated me most, however, was how Chongqing's topography [5] played a role in defining these zones. Because of the city's distinct topography—steep hills, meandering rivers, and verticality—districts and zones were defined by how people moved through the area rather than just by roads or land use. Natural boundaries were established by winding roads, steep staircases, and raised walkways, which gave each area a unique atmosphere. For instance, strolling along the Yangtze or Jialing riverbanks was not the same as walking through the winding streets of a neighbourhood on a hilltop. These physical boundaries exposed the underlying patterns of urban life, and each zone had its own unique set of pleasures and challenges.

This geographic influence was highlighted by the hike up to the Eling Park area, which is situated on a hill with a view of the city. With each zone and district clearly defined by its relationship to the landscape, the city appeared to unfold before my eyes from this vantage point. The districts appeared to be carved into the natural landscape as a result of the rivers cutting the city in half and the hilly terrain creating natural boundaries [8]. Not only were there physical differences between the city's flatlands and higher areas, but there were also socioeconomic ones. I observed how

Fig. 6.4 In one image, there is both a humane scale and a view that suggests this is part of a major metropolis (*Source* Photo is taken by the Author in 2022)

Fig. 6.5 Chaotic in many ways, but also defined, too. Residential areas of Chongqing particular in older areas include boundaries of access, a sense of public against private divisions, defined edges, and connectivity in and around them (*Source* Photos are taken by the Author in September 2020)

Fig. 6.6 The many charming pockets of the city include narrow alleyways between residential buildings, stairways, and everyday living spaces. (Fig. 6.7) (*Source* Photos are taken by the Author in September 2020)

urban planning had adapted to these geographic realities as I strolled through these zones, frequently with unexpected inventiveness. While wide boulevards and elevated roads were constructed in the newer districts to accommodate the city's expanding population, in the older districts, narrow alleys and steep stairways had become the neighbourhood's lifeblood, connecting families and businesses in ways that felt natural.

But the most remarkable thing was how, as I continued to walk, all of these boundaries—historical, geographical, and social—started to meld together. The boundaries between districts were frequently blurred in the more sedate areas of the city, and I could feel how urban life was lived much more fluidly across these zones. People appeared to move freely between zones, engaging with the various facets of urban life, rather than being restricted to the clearly delineated boundaries of their neighbourhoods. It served as a reminder of the city's true vibrancy and interconnectedness. Like a mosaic, Chongqing's zones and districts each add to the overall image of the city's identity [4, 2]. The more I walked around, the more I realized that although these areas were different, they were not separate from one another. The rhythms of everyday life and the people who lived, worked, and walked among them bonded them together. It was not enough to simply observe these zones from a distance, walking through them allowed one to become a part of them and feel the spaces and transitions viscerally.

Fig. 6.7 The city's complex topography often help to create boundaries and connections where needed, separating different areas and zones as well as helping to define where public places begin or end (*Source*: Photo is taken by the Author in 2022)

I found myself thinking about how human agency and urban planning shape these areas as I navigated the city's streets. The streets themselves took on the role of a living map, delineating the city's historical, social, and cultural landmarks in addition to its actual borders. The deeper facets of Chongqing, which are frequently hidden from the untrained eye but come to life when you stroll the streets, were being revealed to me with each step. The physical borders of districts and zones are only one aspect of this section on mapping the city's fabric. It is about how these boundaries, when explored on city walks, uncover the complexities and hidden stories of urban life. The streets and walkways that link Chongqing's districts are much more than just places to travel; they are the arteries and veins that carry the city's lifeblood and tell its history, people, and continuous change. And we can only start to fully comprehend Chongqing's rich tapestry, one district and zone at a time, by strolling these streets (See Figs. 6.8 and 6.9).

Fig. 6.8 Some communities still have a gate that resemble the start of a residential community (*Source* Photo is taken by the Author in 2021)

6.2 The Pulse of Urban Evolution: From Historic Districts to Emerging Zones

As I stepped onto the busy streets of Jiefangbei, the heart of Chongqing, it felt as though the city's pulse was beating in sync with the rhythm of everyday life. In stark contrast to the older neighbourhoods farther out from the city center, this district, with its tall skyscrapers and busy shopping malls, symbolizes the city's rapid modernization. However, beneath the surface of this bustling commercial center, the layers of history were still discernible. I started to realize during my city walks around Chongqing that its zones, districts, and neighbourhoods are more than just places; they are dynamic, living representations of the city's development. And I could sense the city's clear transition from its ancient foundation to its developing future with each step.

Walking through the older parts of Jiefangbei, I began to notice how much of the past still clings to the present. New stores and cafes line the street corners, but there are also old buildings with intricate wooden windows and faded paint. These are relics from Chongqing's colonial past, and as I passed through them, I could feel the stories and presence of the previous residents etched into the stones and bricks. In this case, the change was subtle and integrated into day-to-day existence. The skyscrapers rising in the distance did not seem to bother the locals, who were

Fig. 6.9 The same applies to market communities and public places with gates and signage (*Source* Photo is taken by the Author in August 2024)

sitting on the sidewalks drinking tea and chatting. The contemporary city and the slower-paced, more traditional way of life that had been a part of Chongqing for many generations collided.

However, this disparity was not limited to a single district. I would continue to notice this pattern as I walked through the city. I discovered myself meandering through secret passageways and tiny alleys where time seemed to stop as I moved deeper into Chongqing's historic districts. The history of the city resides in these regions, such as those around the Yangtze River. The shining shopping centers a few blocks away seemed to tell a different story than the dilapidated exteriors of old houses and the incense-filled air. There was a different rhythm to these neighbourhoods, with their private areas and close-knit communities; it was slower and more contemplative, as if the city itself was taking a deep breath before accelerating its rapid growth.

In these older neighbourhoods, I started to understand that Chongqing's history was found in the people's daily lives as much as in the imposing monuments and official structures. It was in the sounds of street musicians performing for onlookers and in the market stalls where merchants offered traditional medicine and fresh produce. These neighbourhoods were bustling with activity and were not remnants of the past. As I walked through them, I could feel how the city's old and new identities were intertwined—how the contemporary metropolis had expanded over, through and around these historic districts while still recognizing their presence.

There was a noticeable change in the city's energy as I made my way from Jiefangbei to the more developing neighbourhoods. The buildings grew taller, the streets widened, and life appeared to move more quickly. I could actually start to see the enormous changes that Chongqing was going through here, in the more recent neighbourhoods and on the outskirts. The more private, historic areas were a far cry from these new areas, which had broad boulevards and gleaming new construction. However, as I strolled, I also started to observe the indications of change: the fading remains of old storefronts being replaced with sleek, modern designs, or the remains of old houses gradually being replaced by modern high-rises.

It was like watching the city expand before my eyes as I moved through these areas. With cranes looming overhead and construction sites on every corner, the streets appeared to be in a state of perpetual change. The city's skyline was changing quickly, but its people, who were its lifeblood, never changed. On the outskirts, I witnessed brand-new communities thriving: families walking down freshly paved streets, kids playing in parks, and businesses opening their doors to greet a fresh generation of locals. It demonstrated how Chongqing was adjusting to the demands of fast urbanization while still making an effort to preserve the interpersonal ties that make the city unique.

However, remnants of the city's past can be discovered even in these new neighbourhoods. The ancient street sellers who had previously set up shop in the city center's winding alleys were now moving their carts to new spots. Additionally, the new apartments that appeared out of nowhere were being occupied by families who had resided in the historic districts for generations. Not only was there a physical transition from the old to the new, but there was also a cultural one. The demands of a modern city were replacing the old rhythms of life. However, the city's core—its residents, its culture, and its lively street life—was preserved throughout the process.

I could not help but notice the conflict between the old and the new as I kept moving through these developing zones. It seemed as though Chongqing was attempting to strike a balance between the demands of a world that was modernizing quickly and the weight of its past. Their once-small, neighbourhood stores had been replaced by global chains, and the streets, which had once been alive with the sounds of pedestrians and horse-drawn carts, were now lined with cars and buses. However, the ancient rhythms persisted, woven into the city's very fabric. I could feel Chongqing struggling with its identity as I strolled through these streets, attempting to embrace the future while clinging to its soul.

The tour of these developing neighbourhoods brought to light Chongqing's urgent need for change while also emphasizing that urban life is more than just the physical infrastructure. It is also about the individuals who live there, give them life, and shape them over time. Even though the city is changing, its future will ultimately be determined by the people who live there, walk its streets, and give its neighbourhoods life [3]. I observed how Chongqing is striking a careful balance between embracing its future and conserving its past while I was strolling around the city. The layers of the city's past and future were revealed to me as I strolled; at times they blended harmoniously, and at other times they stood out sharply. The stories of the people who have lived and worked there for generations are preserved in the alleyways, street corners, and marketplaces.

Additionally, as the city keeps expanding, I have come to understand that creating a cohesive and dynamic urban landscape involves more than just creating new zones. Based on firsthand observations from our city walks, this section provides a profoundly intimate and thoughtful analysis of the historical and current changes to Chongqing's districts and zones. It highlights the dynamic interplay between the old and the new, documenting the city's continuous development and the ways in which the street-level experience contributes to our comprehension of this shift. If you would like any additional changes or additions, please let me know!

6.3 Social Networks and Local Life: Walking Through Chongqing's Zones of Interaction

As I walked through Chongqing with more time spent in places and between places (or as we hear, zones and districts), I started to see that the city's zones and districts are more than just physical locations; they are thriving hubs of social activity where people, families, and communities gather for everyday customs that contribute to the city's identity. Whether I was in Jiefangbei's busy downtown or the more sedate, private residential neighbourhoods on the outskirts, I observed how public areas serve as vibrant hubs for social interaction and provide a window into the social networks that drive the city.

As I walked through the different districts, one of the first things I noticed was how people engaged with their environment. The bustle of business and conversation filled the streets of crowded business districts like Jiefangbei and Nan'an. Individuals moved swiftly, hurrying from shop to shop, negotiating the labyrinth of neon-lit storefronts, their motions coordinating like a meticulously planned dance. However, there was a noticeable sense of social connectedness in spite of the rapid pace. I observed groups of friends huddled outside cafés, laughing and sharing tea, impromptu get-togethers on street corners, and tiny pockets of conversation in the middle of the crowd. These ostensibly insignificant moments were a part of the city's overall rhythm, showing how urban areas develop into social hubs that are more than just places to work or shop.

I became more aware of the nuanced ways that social networks develop in Chongqing's public areas as I kept strolling through these districts. For instance, life seemed to flow naturally from one encounter to the next in the city's busy marketplaces. Families would congregate around food stalls to share meals and stories, and local vendors would engage with frequent patrons, giving them small tokens of appreciation. Despite their apparent insignificance, these exchanges revealed a great deal about Chongqing's social structure [7]. It was at this point that I started to realize how the public areas in these neighbourhoods promote a more profound sense of connection—how residents of Chongqing interact with these areas rather than merely passing through them, creating social networks that are as much a part of the city as its physical infrastructure.

The pace of life slowed as I moved farther into more sedate residential areas, such as those in Chongqing's more outlying districts, but the social fabric remained equally vibrant. The public areas in these more subdued areas of the city were just as significant as those in the commercial districts, despite being less opulent. I observed neighbors introducing themselves, sharing news, and lending a hand with everyday tasks in the tiny plazas and alleyways. These exchanges felt natural, with a sense of trust and camaraderie that was based on enduring bonds. I observed how young parents took their kids to the parks to play and socialize with neighbours, while older men and women gathered at neighbourhood tea houses to share stories about the past.

The more I explored these areas, the more I realized how deeply ingrained social ties are in Chongqing. Social interaction was not merely a consequence of public space; rather, it was the very essence of these residential areas, which were more than just collections of buildings. Even the most ordinary public areas, such as street corners or apartment building entrances, are hubs for social interaction in Chongqing. I found myself lingering in these spaces, observing how people negotiated their place in the urban landscape through conversations, casual greetings, and shared moments. Even in the anonymity of the city, there was a feeling of community here, as though the street itself served as a platform for people to define their own identities.

As I walked through Chongqing's districts, I also started to observe how the various areas of the city mirror broader social patterns. The quick speed of urban growth appeared to cause social disintegration in some areas. I observed that public areas felt more planned and organized in the more recent, contemporary zones. Although the immaculate parks and broad boulevards were lovely, they also seemed impersonal. People were less likely to sit in a park for a long time or to stop and talk to a stranger. These more recent areas' public areas appeared to have been designed more for efficiency and mobility than for interacting with others. In contrast, Chongqing's older, more established neighbourhoods were teeming with busy street corners, narrow alleyways, and outdoor markets—areas that promoted impromptu encounters and fostered the growth of social networks.

This contrast between the older, more natural districts and the newer, more planned areas showed how Chongqing's social life is changing as the city gets more modernized. The intimacy and density of social interaction that define older parts of the city are absent from the newer districts, despite the fact that they might have more formal public spaces. The significance of human-scale spaces—tight-knit communities where people are more likely to engage with one another personally—came to mind as I moved through these zones. Not only do people live and work in these areas, but they also foster the city's social life, foster friendships, and help people make memories.

In one memorable walk into a residential area close to Chongqing's outskirts, I ventured into a place where people lived at a considerably slower pace. The buildings were older, the streets were calmer, and people interacted more casually. Here, I discovered an open-air market where patrons lingered over modest purchases, engaging in conversation with vendors as if they were longtime acquaintances. The sound of laughter reverberated through the streets, and the aroma of street food filled the air. I came to see that social networks in these more sedate areas were based on common practices, such as taking leisurely walks in the neighbourhood parks, going to the market once a week, or having afternoon tea with neighbours. Despite their simplicity, these areas were essential to maintaining Chongqing's sense of community. During these walks, I was particularly struck by how Chongqing's zones and districts serve as a microcosm of the city's broader social dynamics. Public areas are the gathering places where people congregate, interact, and build relationships in every neighbourhood. The city's social life is shaped in these areas, whether it is in the busy business district or the peaceful residential neighbourhoods on the outskirts. Through my walks, I started to realize that these areas of interaction were characterized not only by their physical attributes but also by the people who lived there, the daily interactions that took place there, and the ways in which they reflected the city's social and cultural values.

We are able to learn more about Chongqing's social life and how its streets, alleys, parks, and markets are all interwoven into a complex web of interaction by strolling through these districts. The more we walk through these zones, the more we understood that the real spirit of Chongqing's urban life lies not only in the opulent architecture or the brand-new, gleaming structures, but also in the small, routine exchanges that take place in the city's streets and public areas. I realized during these city walks that every zone and district adds to a broader story about Chongqing's social structure. These places—big and small, private and public, new and old—are what connect the social networks of the city. And the real essence of the city can be discovered in these exchanges, these brief moments of connection.

6.4 Urban Transitions: Walking the Boundaries of Chongqing's Evolving Districts

One thing became more and more obvious to me as I walked through Chongqing's streets: the city is constantly changing, with the lines separating its new and old districts becoming blurred. Walking the streets of Chongqing offers a unique perspective on the city's continuous changes, which are woven together by historical layers, cultural shifts, and rapid development. I witnessed firsthand how the city's borders—whether they be social, cultural, or physical—are constantly changing as I travelled from the busy business districts of Jiefangbei to the more sedate, historic neighbourhoods of Yuzhong.

I had the impression that I was following the development of the city as I walked through the areas that separated the old and new districts. These areas, which combine modern shopping centers with historic marketplaces and high-rise buildings with traditional alleyways, provide a look into the future while clinging to the city's history. For instance, I found myself switching between different environments during one of my walks through the northern Yuzhong District. Across the street were the ruins of a Chongqing that had been left behind: old houses with weathered wooden doors and narrow alleys, while on one side of the street stood a series of new glass-and-steel towers, the pinnacle of modernity, with glossy storefronts and bustling pedestrians. It was easy to see how the old was being merged with the new in this transitional area—a traditional market now next to a posh mall, an old tea house transformed into a hip café. These areas demonstrate the city's adaptability and capacity to incorporate and change its historic areas while preserving a sense of its origins.

The way the city's residents move between these edges with ease was what most impressed me about them. As I watched pedestrians move through these zones, I saw that the lines themselves did not seem to be as clear in the city dwellers' daily lives. Even though the changes were visible on a physical level, it seemed that the people who live in these districts did not notice them as much. The old and the new coexisted in dynamic harmony during this gradual, organic process rather than a sudden change. People moved through these areas, absorbing aspects of both old and new Chongqing as they went, rather than just leaving one zone behind and entering another.

For example, as I strolled around Yuzhong's northern outskirts, I came across a number of transitional areas that represented this dynamic. Both new construction and relics of old Chongqing could be seen in the streets: pedestrian walkways constructed over former roads that had been converted for motorized traffic; small, family-run stores selling traditional goods next to high-end boutiques; and ancient temples hidden behind contemporary office buildings. These streets seemed to be a miniature version of the city's broader changes. In addition to illustrating Chongqing's quick urbanization, the juxtaposition of old and new buildings shows how adaptable the city is while maintaining its cultural and historical significance.

I kept walking as I moved toward the city's outskirts, where it was easier to distinguish between Chongqing's historic neighbourhoods and more contemporary suburban areas. I observed how the city is growing in these areas to accommodate the needs of an increasing population, with smaller, older neighbourhoods and agricultural lands serving as the backdrop for new residential complexes [1]. In some areas, the streets, which had once been winding and narrow, had been enlarged into wide boulevards lined with trees and contemporary transit stations, and high-rise apartment buildings towered over what had once been farmland. By pushing the city outward and creating room for new developments, the building of new roads and transit systems is changing these areas. However, the past is never far behind, even when new structures appear. Even though the growth of contemporary infrastructure was gradually overshadowing these suburban areas, I saw that the historical markers of Chongqing's past—old temples, tiny family-run businesses, and traditional housing—were still very much present.

Through the city walks in Chongqing, we discovered that public space was crucial in mediating between the old and the new in these transitional areas. Large parks and pedestrian walkways had been incorporated into the neighbourhood design in the recently constructed outskirts. These areas functioned as links between the surrounding natural environment and the urban fabric in addition to being places for leisure and recreation. I watched locals use these areas in a number of ways: some went for walks, others worked out in outdoor gyms, and kids played in the designated areas. With their focus on green space and connectivity, these new zones appear to present a vision of Chongqing's future—one that strikes a balance between environmental sustainability and urban growth.

But as I continued to explore the city's older neighbourhoods, I came across a completely different dynamic. Historical neighbourhoods, such as the old parts of Jiefangbei or Ciqikou, had narrow streets that were busy with activity but also far more restricted. These areas' physical borders were less flexible, and the streets appeared more constrained, crowded, and narrow [6]. Smaller public areas were more common here, frequently appearing as street corners or tiny plazas encircled by traditional buildings. These areas are extremely important to Chongqing's communal life, despite being smaller than those in more recent districts. The city's ingrained cultural and social fabric is reflected in the way its citizens engage with these areas, whether through informal discussions, open-air markets, or cultural events. As I strolled through these neighbourhoods, I grew to understand how these historic neighbourhoods still give their residents a sense of continuity and stability in spite of the demands of urban growth.

These walks made it abundantly evident that the lines separating the new and old districts are not rigid but rather are always changing. I was able to observe how Chongqing's identity is shaped by the continuous interaction between various facets of urban history rather than by fixed borders by walking these edges. In addition to being areas of physical change, the transitional zones between the old and new districts are also settings for the negotiation and reshaping of cultural, social, and economic dynamics. These areas are where the city's past and future meet, resulting in a tension that is enlightening and productive. During these city walks, I also came to understand that the boundaries of Chongqing's changing districts are shaped by the experiences and tales of the locals as much as by the actual physical location. I could feel the city's pulse as I moved through these areas—the pulsing rhythms of a place undergoing change, where old customs are being woven into the fabric of the future and new opportunities are continually being created. I gained a deeper understanding of Chongqing's distinct character by walking these boundaries; it is a city that is always changing and evolving while maintaining a strong connection to its historical roots.

References

[1] C Chen L Yu CL Choguill 2020 "Dipiao", Chinese approach to transfer of land development rights: The experiences of Chongqing Land Use Policy 99 104870

[2] Cheshmehzangi A (2020) Identity of cities and city of identities: Examples. In Identity of Cities and City of Identities, Springer Singapore, pp 179–244

[3] Cheshmehzangi A, Tang T (2024) Chongqing: Navigating rapid urbanization and regional development challenges. In: China under construction: Shaping cities through recent urban transformation Springer, Singapore. pp 81–106

[4] Han SS, Wang Y (2001) Chongqing. Cities 18(2):115–125
[5] Jin Y (2022) Urban verticality shaped by a vertical terrain: Lessons from Chongqing, China. Urban planning, 7(4):364–376
[6] Li Y, Cheshmehzangi A, He BJ (2023, November). A framework of urban renewal for leisure space in neighborhood streets: A case study of Shazheng street in Chongqing, China. In: International Conference on Urban Climate, Sustainability and Urban Design Singapore: Springer Nature Singapore, pp 693–705
[7] KF Lim 2014 Spatial egalitarianism as a social 'counter-movement': On socio-economic reforms in Chongqing Econ Soc 43 3 455 493
[8] Y Liu W Yue P Fan Z Zhang J Huang 2017 Assessing the urban environmental quality of mountainous cities: A case study in Chongqing, China Ecol Ind 81 132 145
[9] Mirra F (2022) Urban imaginaries: Contemporary art and urban transformations in China since 2001 (Doctoral dissertation, Birmingham City University).

Part II
The Multi-Identity City

Chapter 7
Urbanity in Flux: The Evolving Essence of Chongqing

7.1 Urbanity and Identity: Defining Chongqing's Multi-Faceted Character

As I started my city walks through Chongqing's streets, I couldn't help but notice how the city's character seemed reflected in the urban landscape. The urbanity of Chongqing is not a monolith like that of many other cities I have been to. It is not limited to one cohesive idea of what a city ought to look like. Rather, it is a multilayered mosaic made of various historical tiers, cultural manifestations, and spatial configurations. This tour of its districts demonstrated how these layers come together to form something distinctively Chongqing: a changing urban fabric that not only embraces its history but also bravely moves forward.

The physicality of the city—its topography, the stark verticality of the urban sprawl—was the first thing that caught my eye. I could feel the geography in my bones as we travelled through various districts, whether it was the busy streets of Jiefangbei or the tiny lanes close to Ciqikou. Built into the mountains, the city rises and falls dramatically, and its streets wind up steep hills, giving the impression that we are walking along the city's spine. Every step I took along one of the main streets in the Yuzhong District made the hill's gradient more noticeable. The expansive expanse of the city below became visible as we climbed higher. Towering structures appeared to be piled on top of one another, a risky yet exquisite illustration of how Chongqing's development is closely linked to its rugged terrain. One of the city's distinctive features is the way it dips and rises [5], sometimes dramatically. It is a setting that not only accepts the topography but also integrates it into the city's overall experience.

During our walks, I observed how the surrounding area changed as we passed through various locations. The streets felt like different worlds, even though they were connected. For instance, Jiefangbei, with its broad, spotless streets lined with upscale stores and skyscraper buildings, was a magnificent representation of

A. Cheshmehzangi, *City Walks in Chongqing*, Urban Sustainability,
https://doi.org/10.1007/978-981-96-6042-1_7

Chongqing's modern ambitions. The region exuded the vitality of a bustling, international metropolis. The allure of Yuzhong District's older neighbourhoods, however, provided a striking contrast just a few blocks away. Chongqing's ancient roots were reflected in its narrow streets and old buildings dotted with small, neighbourhood businesses; daily life and local culture seemed to move more slowly there. This difference was not only caused by the architecture, but also by the atmosphere and the sense that life slowed down as you passed through the layers of the city. Even though they were both in the same city, it felt like traveling back in time to enter these more historic areas from Jiefangbei's glitzy shopping malls. Chongqing is a city at the nexus of history and progress, and this ongoing interplay between the old and the new is what makes its identity so intriguing (See Figs. 7.1 and 7.2).

The fact that Chongqing's urbanity is shaped by how people use the spaces we occupy as well as by the spaces themselves was one of the most profound insights I gained from these walks. Both the more casual settings—streets and alleyways—and the more formal, structured areas—parks and plazas—have a strong sense of community. I observed locals going about their daily lives in places like Ciqikou, whether it was a group of young people congregating around a small street food vendor or an elderly couple conversing while enjoying tea on the porch of a traditional teahouse. These unofficial interactions take place all over the city and are an essential part of Chongqing's urban fabric [6, 1]. People overflow into the streets, utilizing the urban environment as a place to live, interact, and establish a sense of community rather than merely a place to travel through. These exchanges reveal a more general cultural mindset that is ingrained in the city's identity, emphasizing local culture, shared space, and social connection.

As we travelled from one district to another, it became increasingly evident how this complex identity is formed [3]. The city's desire to become a major player on the world stage dictates the design of the newer areas, such as the Liangjiang New Area. Grand boulevards, contemporary buildings, and open areas for transportation were all visible here. However, remnants of Chongqing's past were evident even in these neighbourhoods, whether in the finer points of architectural features that were adapted from traditional styles or in the incorporation of regional cultural allusions into signage and public art [2] (See Fig. 7.3). I had the impression that the city was speaking to me in two languages as I moved through these streets: one that was rooted in its history and customs, and the other that was forward-looking and contemporary. The interplay between these two forces is what gives Chongqing its unique character. It is a city constantly negotiating between its roots and its ambitions.

I was able to observe how Chongqing's urbanity reflects its social and economic diversity by exploring the various districts. Every neighbourhood appeared to represent a distinct facet of the city's overall character. For instance, I discovered that life had a slower, more personal rhythm in Chongqing's historic center. The streets were more constrained, the stores were family-run, and the local way of life appeared to be ingrained in everyday existence. The rich history of the area was reflected in the architecture, cuisine, and language. On the other hand, there was an energy that was difficult to overlook in more recent developments, such as the Yufu Riverfront. The

Fig. 7.1 While Liziba station and its famous metro lines passing through the building is mentioned in other chapters, I believe it is still a sort of urbanity here that suggests more than just a typical functionality that we could see or experience elsewhere (*Source* Photo is taken by the Author in September 2020)

city's desire to modernize and establish itself as a major global city was demonstrated by the quick development of shopping malls, residential towers, and broad boulevards. Every district is infused with this dichotomy between the old and the new, which defines Chongqing's urban character in part. It is a city that recognizes the importance of its history while simultaneously working to forge a fresh, forward-thinking identity. While technologies are left out from this chapter, it is important to note they are forming to be part of the city's characteristics, everyday spaces, and identity (See Figs. 7.4 and 7.6).

I also developed a greater understanding of how Chongqing's past as a gateway between the interior of China and the outside world has influenced its urbanity as a

Fig. 7.2 It may not be entirely visible at first but if you then look closely, you start to see the interplay of places at various levels. You then get to appreciate the interplay between the old and the new is what makes its identity so intriguing (*Source* Photo is taken by the Author in August 2024)

result of these walks. The city has long served as a transit hub, allowing ideas and cultures to coexist. The neighbourhoods that still show the effects of migration and trade are reminders of this history. Stories of how various communities have shaped the city can be found in the streets, marketplaces, and plazas. For example, the old city's Dazhao Temple district narrates the city's lengthy history as a crossroads of culture and religion. Walking through these areas, I could sense the presence of past generations—how the layers of history are still very much alive in the city's urban environment.

Yet, Chongqing's identity has a decidedly modern quality in spite of all these layers. The city's history and rapid development are not the only things that define it. I had the impression that the city was always changing as I strolled through its streets,

Fig. 7.3 Some are traditional representations, some are iconic, and some are meant to create nodes in the city. Signage is a powerful tool in how art and urbanity are brought into urban places (*Source* Photo is taken by the Author in September 2020)

drawing inspiration from its history while envisioning an urban and human-centered future. Chongqing is one of the most distinctive cities I have ever visited because of its urbanity, which is defined by this balance between the old and the new, the traditional and the modern.

I came to understand that the fluidity of Chongqing's identity lies at its core as we strolled through these districts together. It is a city where the past, present, and future all coexist in a harmonious dance; it does not exist in a vacuum. Chongqing's urban fabric is made up of layered histories, spatial forms, and cultural expressions that we experience every time we walk its streets rather than merely seeing from a distance. This dynamic interaction is what gives the city its unique urbanity and strong sense of identity. And as we walked together, I felt a profound connection to

Fig. 7.4 A 24/7 library beside the street where technology meets urbanity and where urbanity meets the urban (*Source* Photo is taken by the Author in October 2019)

this city—an understanding that Chongqing is not just a place, but an experience, one that is constantly evolving and always inviting us to discover more.

7.2 Urban Density and Verticality: Living and Moving in a Three-Dimensional City

As I walked through Chongqing, I couldn't help but be captivated by the city's landscape, which served as a constant reminder that this was no ordinary urban experience. In addition to defining the city's landscape, the city's extreme verticality also defines its identity, rhythm, and how people interact with their environment. Built into steep hills, Chongqing is located in the center of southwestern China. Its streets frequently rise and fall at impossible angles. I could feel the city's energy, movement, and multi-layered complexity as we explored its various neighbourhoods; they were all skillfully woven into this three-dimensional urban structure.

I was immediately struck by the scenery in the center of the city, especially in the Yuzhong District, which is where the city's center is located. No matter where you are, you can't help but notice the overwhelming and thrilling sense of height. The streets of Chongqing test your sense of equilibrium, in contrast to the flat, smooth avenues found in many cities. After traversing narrow alleyways and climbing steep hills, we were abruptly confronted with precipitous drops that gave the impression that the city was constantly rising or falling. I recall a specific stroll close to Jiefangbei, the busy central business district of Chongqing.

Fig. 7.5 A newly installed smart shopping shelves beside the street where, once again, technology meets urbanity and where urbanity meets the urban (*Source* Photo is taken by the Author in 2021)

Modernity and traditional urban forms collide in this area. We were crossing a pedestrian bridge that was elevated above the street below. I felt as though I could see the entire city from this vantage point, including the tall skyscrapers, the sloping roads, the deep gorges that separated the districts, and the far-off rivers that carved through the valley below. It was more than just a view; it was an understanding of how deeply vertical Chongqing is. When we examine how the city functions, this sense of three-dimensionality really comes to light. The conventional perspective on cities often emphasizes horizontal growth, with long, continuous streets and dispersed neighbourhoods. However, Chongqing's streets wind through the mountainous landscape, folding in on themselves and moving vertically in ways that give the city a sense of vitality. Stepping off a steep slope to find another level of the city, climbing narrow stairways to reach an otherwise hidden street, or descending

Fig. 7.6 An example of a popular 3D screen in a busy commercial hub where, once again, technology meets urbanity and where urbanity meets the urban (*Source* Photo is taken by the Author in December 2024)

to find yet another area of the city that feels like it has always existed—despite being invisible from above—all require constant adaptation to the geography. These constant transitions between vertical levels create an ever-changing urban landscape that demands a different way of thinking about urban mobility.

From the steep hills to the underground passages that connect the lower parts of the city, we had both horizontal and vertical experiences of the city. For example, the metro system. It is an essential component of the city's verticality and goes beyond merely being a transportation system. The metro stations and tunnels are frequently carved out of the mountainside, and we had to ride an escalator to reach platforms that appeared to be carved out of the ground itself. These escalators conveyed a lot about how the city adjusts to its terrain because of their sleek, metal surfaces, which stood out against the untamed, rocky mountains. The metro itself, with its extensive, winding routes that stretch widely throughout the city, is a perfect representation of Chongqing's urbanity because it is dynamic rather than linear, extending both horizontally and vertically. This was not a city that could be comprehended by a flat map; rather, it was a place that demanded interaction with its surroundings and allowed you to explore it at your own pace.

This three-dimensional experience is greatly influenced by Chongqing's high-density development. The way the city is crammed into these landscapes, in addition to the steep hills, gives the impression of verticality. Chongqing rises upward in contrast to outlying cities. Public transit and pedestrian walkways are interwoven with residential complexes, office towers, and commercial centers that are stacked on top of each other. The way Chongqing had developed into a multi-layered cityscape fascinated me as we strolled through some of the older neighbourhoods. Some tall buildings are perched on top of what were once residential areas, which are now elevated on multi-story roads and highways, while other buildings are nestled alongside new developments.

The Nan'an District was the site of one walk that especially demonstrated the effects of this high-density development. As we explored, I came to understand that the city's verticality has cultivated a close relationship between its inhabitants and their surroundings, with newer skyscrapers entwined with older residential areas. Streets in some areas of Nan'an narrowed into steep staircases that connected residences, and homes were constructed right into the mountainside. It struck me that residents of these neighbourhoods must have a completely different perspective on the city because they are always going up and down through different levels. The closeness of these areas, where locals coexist with enormous, modern construction, demonstrates Chongqing's capacity to blend the old and the new. It is a city that honours its history while moving forward with plans for the future without hesitation.

The way that verticality affected how people interacted with public space was among the most intriguing things I observed during our walks. Certain neighbourhoods, such as the Ciqikou old town area, had much more pedestrian-focused public areas. The neighbourhood is made up of smaller, older buildings with winding streets and alleys. Even though the streets were steep, they felt approachable, and the open areas between the buildings appeared to serve as gathering places for people. I observed locals interacting with one another informally as they met their neighbours at street corners or strolled to nearby shops. These streets were friendly and human-scale despite their steep inclines, and the intimate scale of the setting permitted these interactions, which felt so natural. However, the feeling of openness and space shifted as we approached more recent areas, especially the high-rise constructions close to the river. The urban density increased here, and the public areas seemed more like platforms made to hold large crowds. It seemed as though the sleek, contemporary buildings and the broad, open roads were made for people to move quickly between locations, negotiating public areas that were intended for efficiency rather than closeness. Nevertheless, there was a fluidity to this movement, a feeling that the city is always changing to accommodate its expanding population.

In these newer zones, urban density had a different quality. The goal was to build layers of urban activity and stack space rather than cram more people into limited areas. It was clear from walking through these neighbourhoods that Chongqing is a city built to support both the upward mobility of its residents and the actual upward mobility of its urban environment. Wide boulevards, metro lines, and elevated walkways all spoke to a city built to move fast, expand, and push boundaries all the time. As I explored Chongqing's vertically layered landscape, what struck me most was how the city's very identity is interwoven with this three-dimensionality. It is a place where social interaction and physical movement coexist in a way that is both beautiful and challenging. Verticality and density are inextricably linked to the rhythms of everyday life in Chongqing, a city that compels you to experience it in multiple dimensions. It became evident to me as I thought back on our walks that Chongqing is defined by this complexity—the interplay of time, space, and movement. It is a city that moves with you, not just one you pass through.

7.3 Public Spaces and Everyday Urban Life: Streets, Markets, and Riverfronts

As I continued my walks through Chongqing, I started to notice and appreciate how much of the city's identity is embedded in its public spaces. These areas are central to Chongqing's social urbanity and are more than just physical spaces. Every public area, whether it be the bustling streets, the serene riverbanks, or the busy markets, reveals something about how people engage with one another and their surroundings. In addition to letting me see how vibrant Chongqing's daily life is, walking through these areas also made it clear to me how important they are in forming the social fabric of the city.

The lively public areas of Chongqing, which are frequently used from early in the morning until late at night, were among the first things that caught my attention. Consider the many markets that are dispersed throughout the city. I came to understand that the market is much more than just a place of business as we strolled through the congested streets of Ciqikou, an old town district renowned for its traditional architecture and marketplaces [4]. It is a venue for social interaction and a place where people can meet, converse, bargain, and mingle. With the markets serving as both a commercial and a social hub, Ciqikou's winding, narrow alleys felt like veins pulsing life into the city.

I came across scenes that seemed firmly rooted in the everyday routines of city life at every turn. Every moment felt like a snapshot of Chongqing's identity as a city centered on its public spaces, whether it was locals haggling for a better deal, vendors selling fresh produce, or elderly citizens sitting on street corners talking about the day's events. In the center of the market, I recall a scene where a number of elderly women were seated at a small tea stall, laughing and chatting while sipping their drinks. The area surrounding them served as a meeting place and a microcosm of Chongqing's complex and dynamic social life, in addition to being a real marketplace.

I now have a better understanding of how Chongqing's urbanity is constructed after exploring these public areas. The areas are not only useful; they are also places where people can interact, connect, and exchange experiences. I was reminded as I walked through the congested lanes that these markets, whether they are modernized ones close to Jiefangbei or traditional ones like those in Ciqikou, are where the formal and informal aspects of urban life come together. They are the backbone of Chongqing's social structure and more than just hubs for trade. Like the city itself, there is a steady yet ever-evolving beat to this place. The riverfront, in contrast to the busy markets, was another kind of public area that made an impression. With the Yangtze and Jialing rivers slicing through the city, Chongqing is a water-shaped city. The riverbanks are essential public areas for the city's citizens. As I strolled along the riverbank, particularly by the Chaotianmen Dock, I noticed how the area is utilized entirely differently than the bustling, cacophonous marketplaces. People came here to escape the intensity of the city by riding bicycles along the river's edge, sitting by the water, or taking leisurely walks along the promenade. The riverfront provided residents with a brief respite and felt like Chongqing's lungs (Fig. 7).

During these walks, it was particularly evident how the riverfront functions as a place for both group gatherings and solitary contemplation. Families were walking by the water, kids were playing by the docks, and couples were holding hands while taking in the peace. Conversely, there were lone individuals sitting on benches, lost in contemplation as they looked out at the river. It seemed to be a place where the city's beat slowed down, promoting introspection and sharing of experiences. Given

Fig. 7.7 Water, waterfront connections, and a new construction site where I believe there is a sense of a banal urbanity *Source* Photo is taken by the Author in 2021

its size and continuous flow, the river appeared to reflect the city's own multi-layered, intricate urbanity—always shifting yet ever-present.

The riverfronts of Chongqing are more than just tourist attractions; they are an integral part of the city's character and provide insight into its social life. The river has always played a significant role in the history of the city, acting as a route for transportation, a representation of wealth, and currently a gathering place for the citizens of Chongqing. During our strolls, I could feel how the locals' relationship with water has influenced how they engage with their surroundings. I was struck not only by the river's beauty but also by how well it was incorporated into the people's everyday lives. The riverbank served as a place for many people to work, socialize, and feel like they belonged in addition to being a place for recreation.

However, I noticed a slight change as we moved into more recent developments and away from the traditional riverfront areas (See Fig. 8). Larger open squares and sleek, contemporary waterfront promenades were features of the more structured public areas in the more recent neighbourhoods. These spaces appeared to be intended for a different kind of interaction, one that is more focused on efficiency, mobility, and visual appeal. I could observe how public areas were changing to meet the demands of a developing, multicultural city as I strolled through these recently built areas, such as the Yuzhong District close to the Chongqing International Convention Center. The areas were bigger, more formal, and frequently more geared toward swiftly and effectively handling sizable crowds.

The largest example of this change was the enormous public square close to the Convention Center, which seemed more like a stage than a place for relaxed strolls. Wide walkways for pedestrians were lined with trees, fountains, and contemporary sculptures, all thoughtfully placed to create a space that draws people in and keeps them organized. Open, spacious, and designed for mobility, the square appeared to represent Chongqing's future—one of modernity, connectivity, and fast urbanization. However, social interaction was still possible in this more modern setting. Elderly residents observed from nearby benches as groups of young people congregated by the fountains, chatting and laughing. The more structured, modern spaces were still places where people could come together, albeit in a different way. During these strolls through Chongqing's public areas, I was particularly struck by how the city's identity is influenced by both the built environment and how its citizens use it. People's interactions with their surroundings define Chongqing, and the way its citizens use these public areas reveals a lot about the social urbanity of the city. Every public area in Chongqing, from the lively Ciqikou markets to the serene riverbanks and contemporary squares, has a distinct personality. Through their layout, function, and ambiance, these areas act as windows into the city's development, reflecting its past, present, and future.

As I walked through these areas, I came to understand that Chongqing's public life is characterized by social interactions that occur there as well as the city's architectural features. These areas, where the intricate facets of history, culture, and daily life converge, are where the city's urbanity really comes to life. As I continued to walk, I came to realize that these public areas are more than just the backdrop for the city's events; they are the beating heart of Chongqing's social life, constantly changing,

Fig. 7.8 Here, I expect to be substantial development-based changes soon, if not already in place. A reminder of the past, a reminder of the history of industrial areas across the city remain to resemble the city's urbanity in waterfront edges. *Source* Photo is taken by the Author in September 2020

constantly in motion, and constantly bringing people together in both visible and invisible ways.

7.4 The Future of Urbanity: Continuity, Change, and Emerging Narratives

Chongqing has a rhythm that is hard to describe but impossible to overlook. It is the rhythm of the river crashing into Chaotianmen's piers, the sound of footsteps on the steep staircases that wind through the city's rolling landscape, and the steady hum of the cable car as it crosses the Jialing River. The city is alive with contradictions: ancient and modern, permanent and ephemeral, chaotic and purposeful. Experiencing this dichotomy firsthand while strolling around Chongqing allows one to simultaneously feel the pull of history and the thrust of modernity. And these walks give me the most insightful looks into the city's identity as I follow its changing urban character.

We walked through the revitalized urban areas of the former industrial belt on one of our most moving city walks. These places, which were formerly rough and functional, have evolved into thriving hubs for community involvement and artistic expression. The street in Huangjueping feels like a living canvas, a synthesis of the artistic energy of the future and the industrial residue of the past, with murals covering entire buildings. The same that you can now see across the city where places become colourful and often meaningful, too (See Fig. 9). The contrast is remarkable: former factories have been transformed into co-working spaces and cultural centers, while old warehouses are now home to independent galleries. Rather than conflicting, continuity and change are interacting to create a new urbanity where the past influences the present without limiting creativity.

As we continue our strolls, we come to the center of Ciqikou, where every cobblestone and alleyway still bears the imprint of history. The aroma of freshly roasted tea leaves, the cries of street vendors, and the soft murmur of tourists running their fingers along the weathered wooden beams of old buildings fill the winding streets. It is a location that asserts its identity with a unique sense of place, defying the homogenization of contemporary urban landscapes. Even so, the city is changing—digital payment systems make transactions in century-old teahouses easier, and tiny contemporary cafés peek out from behind traditional storefronts. The delicate yet constant balance between adaptation and preservation supports Chongqing's ability to smoothly integrate its past into its present.

We travel through Jiangbeizui, where Chongqing's urban ambition is most evident, as the city expands its skyline with ambitious new developments. As symbols of a city focused on the future, the streamlined glass towers represent the meeting point of the Jialing and Yangtze Rivers. However, when we look past the glittering exteriors, we can see the subtle ways that urbanity is changing. These developments now incorporate public spaces, which were previously considered an afterthought in high-density cities. Examples include waterfront promenades that encourage people to stay rather than just pass by, and open-air plazas where office workers and street performers interact. The city's expansion must be inclusive, and in its aspirations, it must preserve the closeness of human-scale interactions that characterize a city that is lived in.

Our walks also take us through the labyrinthine lanes of Shancheng Xiang, an old neighbourhood seemingly untouched by time yet framed by the encroaching modernity of surrounding high-rises. In this neighbourhood, small restaurants serve hand-pulled noodles that have not changed in decades, and senior citizens still play chess in communal courtyards. Nevertheless, we notice the slow changes with every visit—demolishment sites here, renovations there—a reminder that Chongqing's continuity is never static. Can these spaces' souls survive in the face of unrelenting modernization? is the question that remains. As we stroll through these streets, we come to understand that Chongqing's urbanity is characterized by its ability to adapt to the changing tides of time rather than by any one era.

Fig. 7.9 There are many ways to see "*Welcome to Chong Qing!*" across the city, and here is an example that is artistic, contemporary, and a representative of the city's culture and urbanity. *Source* Photo is taken by the Author in August 2024

Standing on Raffles City's rooftop and looking down at the very streets we have walked is possibly one of the most telling parts of our city walks. The city spreads out in front of us from this angle, a dynamic, multi-layered, and dense structure. The neon signs flicker as day gives way to night, the bridges connect its various parts, and the rivers meander through its center. It conveys a sense of movement and a city that is both dynamic and firmly rooted in its own changing identity. In this place, history is not lost but rather transformed into something new, and urbanity is not only about the constructed environment but also about the experiences, stories, and lives that give its streets life.

As we think back on our multiple walks in and through Chongqing, it is evident that the city's future depends on how people continue to occupy, traverse, and reimagine its spaces rather than solely on tall skyscrapers or ambitious infrastructure projects. The urbanity of Chongqing, a city that defies easy categorization, is a negotiation between its rocky past and its vast future. And we carry with us the echoes of the past and the hints of the future as we walk its streets one last time—a city that changes, evolves, and, most importantly, endures.

References

[1] M Bonato 2024 Semiotic ideology and mutable sense of place: Chinese ecological urban renewal through the lens of advertising codes Polit Geogr 115 103232

[2] Chen J, Hu W (2024) Geo-semiotic analysis of shared streets in urban historical districts: The case of Jiefangbei, Chongqing, China. Land, 13(8):1232

[3] A Cheshmehzangi 2015 Urban identity as a global phenomenon: Hybridity and contextualization of urban identities in the social environment J Hum Behav Soc Environ 25 5 391 406

[4] Cheshmehzangi A, Tang T (2024) Chongqing: Navigating rapid urbanization and regional development challenges. In: China under construction: shaping cities through recent urban transformation. Springer, Singapore. pp 81–106

[5] Larson C (2010) Chicago on the Yangtze: Welcome to Chongqing, the biggest city you've never heard of. Foreign policy magazine. Available online from: http://faculty.washington.edu/momara/Chicago%20on%20the%20Yangtze.pdf

[6] Xu Q (2017) The challenge of place identity in the making and experience of new Chinese urban square (Doctoral dissertation, University of Sheffield)

Chapter 8
Urban Life Rhythms: Mapping Chongqing Through Its Streets

8.1 The Flow of Everyday Life: Walking Through Chongqing's Urban Pulse

Chongqing has a pulse that runs through its alleys, streets, and riverbanks, influencing the daily lives of its residents. The movement of street sellers erecting their booths before dawn, the ceaseless ebb and flow of labourers traversing the city's steep topography, and the subdued hum of life taking place in obscure nooks all contribute to this rhythm, which is not distinct but rather rich, multifaceted, and dynamic. It is possible to experience the various identities that coexist in this city of steep hills and meandering rivers [2] by walking through Chongqing, which is an immersion into this flow.

In Chongqing, the morning is a spectacle unto itself. The energy is evident in the wet markets, where it all starts. The aroma of freshly chopped herbs, the salty tang of fish tanks, and the crisp scent of steaming baozi filled the air as we made our way through the winding streets of Ciqikou's morning market. Every day, vendors announced their specialties, negotiating with elderly women who examined every vegetable with the attention to detail of experienced chefs. Urban life is quite tactile here, with exchanges, gestures, and agreements creating an unspoken language between the buyer and seller [5]. These markets serve as social gathering places as well as places for business, where discussions about the best ways to cook a dish or the rising cost of pork are woven into the daily routine. The same sort of characteristics that is replicated in Hongya Cave Cultural Center (See Fig. 8.1).

The hustle and bustle of workers fills the streets outside the marketplaces. There are narrow walking routes through the markets as well as quite little pockets of spaces where people could rest, eat or drink, and socialize (See Figs. 8.2 and 8.3). A mixture of students, street vendors, and white-collar workers travel between areas on the city's vital transit systems, the buses and monorails. During one of our strolls, we passed the Liziba monorail station, where the train famously passes through a residential structure. This bizarre scene perfectly depicts the city's dynamic and dense urban

A. Cheshmehzangi, *City Walks in Chongqing*, Urban Sustainability,
https://doi.org/10.1007/978-981-96-6042-1_8

Fig. 8.1 Hongya Cave Cultural Center is a major tourist attraction point where you can see businesses, urban life, social gatherings, food, and culture—all in one place. (*Source* Panorama photo is taken by the Author in October 2019)

environment. It was evident that the city's complex identity was mirrored in these everyday changes as commuters entered and exited with ease, each engrossed in their own world of ideas and destinations [4]. These areas are briefly occupied by people from various backgrounds who speak a mix of Mandarin and Chongqing dialect before vanishing into the maze of underground passageways and high-rises.

Then there is lunchtime, revealing yet another rhythm of Chongqing's urban life. The pace changes with the lunch break, when throngs overflow onto sidewalks dotted with hotpot restaurants and noodle stalls. We strolled along Shancheng Lane, a historic street that becomes a sea of temporary tables and plastic seats, with folks huddled over bowls of spicy Xiaomian. Here, the city's gastronomic character really

Fig. 8.2 Linear and narrow walking routes vs. socialising quite places in one location as part of Ciqikou old town, one of the regenerated old towns in Chongqing. (*Source* Photos are taken by the Author in October 2019)

Fig. 8.3 Ciqikou old town is a regeneration area where new cafes, restaurants, and socialising spaces are major parts of this vibrant zone, (*Source* Photos are taken by the Author in October 2019)

shines. A symphony of midday life is created by the rhythmic cutting of vegetables, the clatter of chopsticks against bowls, and the air infused with spices. Eating here is casual and communal, in contrast to other cities' regimented dining cultures, which furthers Chongqing's image as a place where food is both a social event and a necessity.

In the afternoon, the city's heartbeat shifts to one of movement, commerce, and building. The contrast between the old and the new was evident to us during our stroll through Guanyinqiao's commercial district, where tall shopping centers coexist with teahouses that have been in the family for many generations. Delivery drivers on e-bikes sped through the traffic, representing the fast-paced efficiency of contemporary urban life, while hard-hatted workers had breaks along the sidewalks, drinking tea from thermoses. On the other hand, we discovered older guys playing chess in one of the more peaceful lanes just off the main road. Their presence served as a reminder that, despite Chongqing's rapid development, remnants of a slower, more traditional pace may still be found.

The streets assume yet another identity as nightfall draws near, which reflects on Friedmann's study [6] of place and place-making in Chinese cities. Along the banks of the Jialing and Yangtze Rivers, where the city's pulse slows, if only briefly, the change from day to night is most noticeable. As we strolled down Nanbin Road, we saw couples walking hand in hand, elderly people doing slow tai chi, and young artists drawing the skyline while neon lights flickered. In many respects, the riverside serves as a barrier separating the city's more contemplative, poetic side from its unrelenting pace. The river, which is ever-present and strong, provides a place to stop and reflects the energy of the city. However, nightlife explodes in other places as fast as the calm settles in others. The commercial center of the city, Jiefangbei, comes alive with energy. It was impossible to overlook how the city adjusts to the demands of its citizens as one strolled through its lit streets, past crowds of youths posing for pictures and elderly folks seated outside mahjong parlours. With every hour that goes by, Chongqing changes, evolves, and reinvents itself; it never sleeps.

These everyday routines are ingrained with Chongqing's multi-identity culture. It is a city where modernism and tradition, rapid and leisurely, and old and new coexist. The encounters, trades, and common areas that define the character of the city are what give urban life its pulse, not merely movement. More than merely observing, strolling through Chongqing's streets is an act of involvement that allows one to sense the city's multiple identities in flux. Every step unveils a new layer, a new aspect of the city, and it is inside this layering that Chongqing's actual character emerges. Every street and lane in the city tell a story about its inhabitants as it moves, breathes, and pulses. Traveling through its metropolitan center is a voyage not only through geography but also through time, culture, and the always changing character of a city that defies easy categorization.

8.2 The Social Fabric of the Streets: Encounters, Rituals, and Public Spaces

You can experience life in its purest form when you walk around the streets of Chongqing. More than just locations to get around, the city's streets, alleyways, and open areas are hubs for human interaction where social customs are carried out every day in a smooth rhythm of movement, trade, and tradition. The interactions between vendors, locals, and tourists create the unique texture of urban life in each district. We became fully immersed in these areas throughout our daily walks, observing how routine and spontaneity coexist in public life.

We were in the center of a large street market in Yangjiaping, where business and community coexisted peacefully. With their carts brimming with home items, fresh veggies, and a variety of street food, vendors beckoned to prospective buyers in the Chongqing dialect's sing-song rhythm. Young moms haggled over the cost of fruit with the keen accuracy of people who had done it a thousand times before, while elderly men sat on low stools and smoked long pipes while having animated conversations. These markets serve as much than just commercial centers; they are also microcosms of Chongqing's urban life, gathering places where people gradually become acquainted.

Later on, the social fabric in Shapingba's university district changed colour. Students crowded the streets here, some rushing from one class to another, others loitering at roadside bookshops or drinking bubble tea outside tiny cafés. A young curiosity and engagement with the world was evident in the intellectual vitality of the conversations held here. The formality of the adjoining streets stood in stark contrast to the regimented college setting. We stopped at a tiny noodle shop where groups of students congregated to talk about anything from current events to exams, their laughter blending with the sound of hot woks being splattered by chili oil. These interactions addressed the city's changing character, where knowledge and day-to-day existence are effortlessly woven together, and tradition meets contemporary.

We observed how public areas are transformed into social stages while strolling through Nan'an's older neighbourhoods. A group of pensioners had met for an afternoon game of mahjong in a little area between old apartment complexes. Laughter and good-natured joking were interspersed with the sound of tiles clattering on wooden tables. A woman nearby used a brush soaked in water to practice calligraphy on the sidewalk, reflecting the fleeting aspect of street life with her work. In the face of Chongqing's constantly shifting urban landscape, these social activities—which are carried out every day in innumerable locations throughout the city—act as pillars of familiarity, maintaining social links.

Public places take on a different personality at night (Fig. 8.4). We were in the thick of a vibrant dance party in the pedestrian streets of Guanyinqiao, where middle-aged ladies dressed in vibrant costumes were moving to the beat of popular music. The dancing groups, which are a common sight in Chongqing, serve as a means for individuals to reclaim urban spaces as their own and are more than just a recreational activity [3]. Families and groups of friends stopped to speak and munch at the food vendors around them, which sold grilled corn and skewers. Here, the city's shared experience culture is strengthened as the streets are transformed into festive spaces.

The contradictions of Chongqing's social milieu were most noticeable during our strolls around the busy commercial corridors of Jiefangbei (Fig. 8.5). We observed old people selling handcrafted souvenirs on street corners amid upscale stores and tall skyscrapers, serving as a reminder of the city's multi-tiered economy. An unexpected soundtrack to the modernity around him was created by the mournful strains of an erhu played by a street musician nearby. The social fabric of the city is defined by these contrasts: the past and the present, luxury and modesty, structure and spontaneity. It was impossible not to be moved by the resilience and adaptability of those who inhabit them.

The mobility of Chongqing's public areas is what distinguishes them. People who utilize them are always redefining them; they are influenced by customs but are receptive to fresh ways of expressing urban life. The streets of Chongqing are teeming with stories just waiting to be unearthed, whether it be an unplanned performance, a late-night get-together, or a private discussion between strangers. As we moved through various areas, we did more than simply observe the city; we were knitted into its social fabric by the innumerable encounters that comprise everyday life.

Fig. 8.4 The sign 'One Night in Street' truly resembles what urban life means in Chongqing. (*Source* Photo is taken by the Author in August 2024)

Fig. 8.5 The busy life of commercial corridors of Jiefangbei are represented well in this public life. This moment is captured after the midnight hours when the city is still alive, but perhaps not as busy as other times. (*Source* Photo is taken by the Author in August 2024)

8.3 Work, Hustle, and Mobility: Navigating the City's Economic Heartbeat

The first thing that caught my attention during our city walks through Chongqing's crowded streets and avenues was the tangible energy that makes the city unique. From the construction cranes that dot the skyline to the constant stream of people working through the streets, marketplaces, and offices, Chongqing is more than simply a location—it is a living thing that is always in motion. Every step you take in this city reveals a different aspect of its multi-identity character and economic pulse, where identities, cultures, and vocations all fit in perfectly with the urban fabric.

We started our walk near the iconic Jiefangbei district, where towering skyscrapers stretch towards the sky. It was simple to picture the city's tall skyscrapers teeming with government, tech, and finance experts. The energy of the commuters and employees rushing through the city, their movements coordinating the urban beat, was what really drew my attention, not simply the glass facades. Everyone plays a part in this special ballet of labor and movement that drives the city ahead, from the street vendors selling freshly prepared noodles to the office workers in business suits.

Chongqing's economy thrives on diversity. I observed how the disparities between various occupations shaped the urban environment as we strolled around the streets. We descended to the docks by the river, where the movement of cargo was as steady as the Yangtze River. There, cranes appeared to be as alive as the people who operated them, and dock workers carefully unloaded cargo. This harsh yet lovely contrast served as a reminder that no occupation is unimportant in a metropolis like Chongqing. Once a center for grain and coal, the bustling docks are today alive with international trade and commerce. The inhabitants of Chongqing also leave their mark here, negotiating the economic and literal currents of the river.

We ventured into the city's underground shopping tunnels from the docks. These areas, which are frequently disregarded, demonstrate the creativity of the city's working class. We were surrounded by consumers, store employees, and merchants as we made our way down into the chilly, barely lighted hallways; each had their own set of responsibilities. Beneath the glistening city above, these underground passageways function as the urban economy's veins, transporting goods, services, and the livelihoods of innumerable people. Here, I was able to experience Chongqing's economy at its most basic level—the cadence of labor that was very different from the opulence of high-rise buildings.

Unquestionably, there is a conflict between the old and the modern notwithstanding the city's phenomenal economic progress. Our stroll down the streets close to the city's industrial areas, where warehouses and factories served as a sharp reminder of the area's industrial past, made this clear. I observed employees—many of them young, some even older—as we went by, moving bulky objects, modifying machinery, or putting things together. Even though they are less noticeable in the modern digital economy, the industries are nevertheless an essential component of Chongqing. Their presence illustrates the city's dual identity, with one foot firmly rooted in its industrial heritage and the other moving forward into a future characterized by innovation, technology, and services.

The disparities of Chongqing's economic landscape were vividly brought to life by the experience of traversing these varied areas. People are migrating from building sites to offices, from riverbank piers to retail avenues, from industrial areas to the opulent downtown high-rises, and there's an overwhelming sense of mobility everywhere you look. Every profession in the city adds to the larger economic machine, and there is always a bustle. This continuous mobility reflects the city's cultural fluidity as well as its economic vitality. People from different cultures, occupations, and backgrounds come together in Chongqing and live side by side, each establishing their own niche in the vast urban environment.

As we continued our walk, I noticed how urban life here is defined not just by where people work, but by how they move through the city. In this context, mobility encompasses more than just transportation; it also refers to opportunities and the flexibility to move between various areas of life and work. Mobility in Chongqing is closely linked to its economic structure, as demonstrated by the office worker's daily travel and the dock worker's quick motions when loading containers. The vital link between these diverse areas is the city's public transportation system, which includes a vast network of buses, metro lines, and even cable cars. It is not uncommon to see a young professional and a factory worker traveling together on a train on any given day, both of them equally committed in the development and mobility of the city.

However, the rhythms of work that reverberate throughout the city are what really define Chongqing's work life, not just the actual locations. The urban tempo is influenced by the varying shifts of workers, the fast-paced nature of business meetings, and the gradual yet consistent rhythms of commerce. The city thrives on its work, hustle, and mobility, which can be seen in everything from the peaceful moments of contemplation in its temples to the cacophony of street merchants at night.

The more I went on these walks, the more I understood that Chongqing is a multi-identity city due to the variety of economic activities that influence the city's scenery as well as the diversity of people who live and work here. Every area of the city, from the tall office buildings to the underground shopping plazas, has an own personality, rhythm, and profession to tell. And as we strolled through the streets, I realized that the dynamic interactions between all of these identities and professions characterize Chongqing's urban life rather than any one of them. Every individual we encountered and every area we entered contributed to the greater fabric of this dynamic, diverse metropolis.

8.4 Urban Life in Transition: Continuity, Change, and Future Possibilities

It became evident to us as we walked through Chongqing that this is a city that is always changing and reinventing itself while maintaining a strong connection to its history. Its vibrant urban life is simultaneously anchored in tradition and forward-thinking, fusing the two in ways that provide a singular urban experience. This last part of our trip together in Chongqing is about thinking back on how urban life is changing in the city—how the past is not gone but is changing, and how contemporary changes are changing how people live, congregate, and interact with the city.

Our city walks took us through the narrow alleyways of Ciqikou, where the old Chongqing still exists in the form of traditional tea houses, local markets, and street vendors. I could just feel the weight of history as I strolled around these old streets. The aroma of freshly made dumplings blended with the subtle incense from a neighbouring temple, fostering an atmosphere where customs from the past continue to play an important role in day-to-day activities. However, Chongqing is not stuck in a bygone era, as evidenced by the existence of contemporary cafes, gift shops, and even the sound of mobile phones ringing. The city is continuously balancing its history and future. In this city, the fast-paced rhythms of modernization and international trade were transforming the customs of storytelling and tea drinking.

I became more aware of how Chongqing's urban life was evolving as we continued to explore these old areas. The modern, sleek designs of new high-rises coexisted with the preservation of historic buildings. Nowhere was this more apparent than during our stroll along the Yangtze River's shore. Even while the city's industrial heritage is never far away, it is quickly being surpassed by new constructions, as was evident when standing by the riverbank. What may have been factories and warehouses are now being converted into chic cafes, luxurious residences on the river, and public areas for pleasure and relaxation. However, the river itself—the city's lifeblood—remains constant despite all this upheaval. It is a living example of both change and continuity.

I was astounded by the difference between the older residential sections and the high-tech zones as we strolled around several districts. We explored Chongqing's burgeoning IT areas in the city center, where innovation was evident everywhere we looked. Young professionals occupied the modern glass buildings of IT companies, co-working spaces, and innovation hubs, which stood in stark contrast to the modest ancient districts we had just passed. Even the streets themselves demonstrated this ongoing change and the way that technology is permeating every aspect of daily life. AI-enabled public services, smart trash cans, and interactive street signs are all examples of how public areas are being redesigned as digital ecosystems. The city's adoption of technology was changing how people move, engage, and experience public life in addition to how they work.

Nonetheless, there is an indisputable conflict between these two forces: the continued popularity of traditional ways of living and the quick industrialization of metropolitan areas. I observed how many of the older residential neighbourhoods

Fig. 8.6 A relatively new residential compound is raised over the street level with many new public areas and communal spaces for commercial and leisure uses, but mainly restaurants, takeaway shops, and supermarkets. Here is an example of a new community that has kept traces of the past through its pulse of urban life. (*Source* Photo is taken by the Author in September 2020)

are gradually being replaced by contemporary apartment complexes as I strolled through them (See Fig. 8.6). However, the ancient cycles of existence continue to exist here as well. Under the shade of ancient trees, the elderly still congregate in tiny courtyards to play mahjong or converse. While their parents shop at the street markets, children play traditional games and run through tiny alleys. These everyday connections and community customs are still essential to the locals' identity, even in the face of the soaring towers around them. I saw the best illustration of continuity in the face of change here.

The adaptation of Chongqing's urban life was among the most remarkable aspects of our walks. Many aspects of the city's inhabitants' movement and social interactions have changed, yet certain elements have not changed. For example, although Chongqing's traditional street markets remain an important part of daily life, their layout is evolving. There is a new sense of organization even though the commotion of negotiating and bargaining is still present. The marketplaces are now hubs for food delivery services, and digital payments have supplanted cash payments. Even as technology and contemporary trade influence how people perceive these areas, the market culture's vitality and vigour are being maintained.

Our visit to one of the expanding public parks in Chongqing served as another illustration of this change. I was intrigued by how the city is redefining the idea of public space as we strolled through the park. While traditional Chinese parks and gardens emphasize nature and peaceful contemplation, Chongqing's more recent parks are evolving into places for socializing, sports, and activities. Whether they were exercising, playing games, or simply lounging in the sun, young families, students, and retirees all did so with a decidedly modern sense of connectedness to the city and to one another. These areas are now hubs for engagement, not only with the natural world but also with contemporary urban life, technology, and communities (See Figs. 8.7, 8.8, and 8.9).

The fact that Chongqing's future is not being constructed in a vacuum from its past is what most intrigued me about our walks. The city's expansion is a reimagining of its traditions rather than a rejection of them. A palpable feeling of possibility

Fig. 8.7 Often, urban life is about public realm and public interactions that are there for us to observe and experience in a moment. There could be continuity of such moments and places, and there could be just a singular moment of interaction in a sort of urban life that is captured here from two views. (*Source* Photos are taken by the Author in September 2020)

permeates the atmosphere. We could see that Chongqing is a city in change [1], but one that is doing so in a way that respects its distinct identity as we stood atop one of the city's numerous hilltops and gazed out over the expansive metropolitan landscape. Instead of creating conflict, the contrast between the old and the new is a discourse about the city's past and future. This sense of transition, of possibilities ahead, was everywhere in the city, whether it was in the smart infrastructure projects we saw, the new areas that were being developed for creative industries, or even the increasing number of neighbourhood-based urban projects. Like the city itself, I had the impression that Chongqing's residents were redefining their identities as they moved forward, bringing their customs and experiences with them. The future is full of possibilities in a city where work, mobility, and change are integral parts of daily life. It is a future in which new places and ancient ways, technology and tradition, combine to reshape what it means to be a Chongqing resident.

Our walks helped me realize that Chongqing's urban life is about appreciating the history while managing the swift changes that will determine its future, not about running away from it. As we experienced the city's ever-changing urban identity together, I felt a strong connection to its pulse—its continuity, its change, and its limitless potential.

Fig. 8.8 From the top view, you could see a contemporary installation art, urban realm, walking areas, and visible urban life on the ground. (*Source* Photos are taken by the Author in September 2020)

Fig. 8.9 A main public place during the COVID-19 pandemic, which still served as a place of interaction, meeting, and socialisation. (*Source* Photo is taken by the Author in 2021)

References

1. Bao HX, Li L, Lizieri C (2019) City profile: chongqing (1997–2017). Cities 94:161–171
2. Cao Y, Vazhayil, AM, Wantzen KM (2023) River culture in China and India, a comparative perspective on its origins, challenges, and applications. In: Wantzen Karl M (ed) River Cultures–Life as a Dance to the Rhythm of the Waters, pp 281–312

3. Chen L (2022) Study on the relationship between spatial feature and usage of alley space in urban settlement: A Comparative Methodology in Case of Xing'anli, Shanghai and Shibanpo, Chongqing. Doctoral dissertation, Waseda University
4. Cheshmehzangi A (2020) Reflecting on urban identities. In: Identity of cities and city of identities, Springer, Singapore, pp 133–178
5. Cheshmehzangi A (2023) Regenerating the local characteristics: the impacts on city life experience and identity of Shangri-La's old town. In: Mapping urban regeneration: city life experiences in Yunnan, China. Springer Nature Singapore, Singapore, pp 219–237
6. Friedmann J (2007) Reflections on place and place-making in the cities of China. Int J Urban Reg Res 31(2):257–279

Chapter 9
Diversity in Urban Fusion: Blended Identities of Chongqing

9.1 Walking Through Layers of Identity: A City of Many Belongings

Chongqing is a city of many identities—each shaped by waves of migration, shifting economies, and the constant negotiation between the old and the new [6]. I have always had the impression that this city is a location where several histories, cultures, and lifestyles cohabit rather than having a single identity when I walk its streets. The city is diverse not only in terms of population but also in terms of architecture, space [1], and experience (See Figs. 9.1, 9.2, 9.3, 9.4, and 9.5). It is seen in how various neighbourhoods embrace the future while clinging to their history, how individuals from all backgrounds coexist in the same metropolitan areas, and how new generations retain and reinterpret traditions.

We visited locations where the city's various identities converge and blend to comprehend its variety. We went on one of the most eye-opening treks through Ciqikou, a place that has long been a center of migration and change. It is more than just a well-preserved old town; it is a place where various facets of Chongqing's identity can be tracked in its streets, buildings, and day-to-day activities. It is frequently referred to as a "living museum". Our tour of Ciqikou was more than just a leisurely stroll through a charming historical place; it was an exploration of the changes in Chongqing's identity over the ages [4]. The contrast between the old and the new was immediately noticeable at the main entrance, where we began. We were reminded that this area was constantly changing by the contemporary coffee shops and craft beer bars tucked amongst traditional tea houses, even if the old streets, which were lined with wooden structures from the Ming and Qing eras, appeared to be stuck in time.

Ciqikou has long been a cross-cultural meeting place because of its past as a port town and commercial center. As we explored more, we discovered the effects of migration on the region. Here, generations of workers, craftspeople, and traders from across Sichuan and beyond had made their home, incorporating their customs

A. Cheshmehzangi, *City Walks in Chongqing*, Urban Sustainability,
https://doi.org/10.1007/978-981-96-6042-1_9

Fig. 9.1 Diversity in architecture in central Jiefangbei, which includes one of the most famous iconic buildings of the city known as the Chongqing Clock Tower for the first time visitors, originally as the War Victory Monument and People's Liberation Monument, previously known as the War Victory Monument, which was then renamed to the Chongqing People's Liberation Monument after 1949. (*Source* Photo is taken by the Author in October 2019)

Fig. 9.2 Diversity in architecture is a unique characteristic of Chongqing as you walk around different zones or districts. (*Source* Photo is taken by the Author in 2021)

Fig. 9.3 Diversity in social activities and socialising spaces of Hongya Cave Cultural Center. (*Source* Photo is taken by the Author in October 2019)

Fig. 9.4 Diversity in public places is experienced throughout the city, where there are opportunities for interaction as well as places of threshold and connections between shopping malls and different zones. (*Source* Photo is taken by the Author in February 2023)

into the city's fabric. We were reminded of this lengthy history of migration and colonization by the languages we heard on the streets [8], some of which were distinctively Chongqing and others of which had influences from nearby provinces. While we were there, an elderly couple was making freshly prepared 'hongyou chaoshou' (translated to red oil wontons) at a small, family-run restaurant. They described their origins to us, following their family's path from rural Sichuan to the center of Chongqing's business areas. Over the course of several generations, the

Fig. 9.5 Diversity in zones, where we can see the connection between the new commercial zone and one of the old (mainly residential) communities in Chongqing. (*Source* Photo is taken by the Author in August 2024, under the extreme summer heat of 48 degrees Celsius)

city's character was shaped and reshaped by the experiences of numerous individuals who had arrived in search of opportunity.

However, a younger generation was redefining what it meant to be a Chongqing resident just a few doors down. We went inside a store that sold modern designs based on traditional Shu embroidery. The designers were a group of young people who had studied elsewhere and came back to incorporate international style into their local handicraft. The essence of a city that views variety as a distinguishing feature rather than a problem was captured by this mingling of influences—local and global, ancient and new (Figs. 9.6 and 9.7).

During our walking tour throughout the visits, I was particularly impressed by Ciqikou's transformation from a center for working-class immigrants to a cultural icon. Once a place of economic survival, where dockworkers and traders fought to make a name for themselves, it was now a destination for cultural tourism, where history was carefully chosen for tourists. Although it had taken on new forms, the diversity of the past had not vanished [5]. This change was especially evident in the local cuisine culture. Families that had lived in the area for many years continued to operate the traditional kiosks selling homemade tofu pudding and peanut-stuffed tangyuan (glutinous rice balls). However, they were accompanied by a flurry of new culinary experimentation, such as artisan coffee flavoured with regional ingredients,

Fig. 9.6 Here is a local look. The preserved old is often next to the newly developed areas of the city. (*Source* Photo is taken by the Author in September 2020)

Fig. 9.7 Here is a global look. The new is usually influenced by various architectural styles, some that appear to be global or contemporary/modern. (*Source* Photo is taken by the Author in 2021)

fusion desserts made with Sichuan peppercorns, and even worldwide food modified to suit Chongqing's spicy palette.

An significant question was brought up by this evolution: does urban development always maintain a place's authenticity or does it substitute one identity for another? Economic gains were certainly a result of Ciqikou's rebirth as a cultural icon, but the city's performance and consumption of identity were also altered. Once gathering spots for residents to exchange stories, the traditional tea rooms were now primarily used by visitors. But even with this change, there were still places where the past refused to be completely changed. We found an antique calligraphy store in a little lane off the main tourist route. The owner had been teaching traditional Chinese ink painting for decades, unaffected by commercialization.

The way that multi-identity manifests itself in daily life, rather than merely the historical change, was what made our walk in Ciqikou so eye-opening. In a single afternoon, we witnessed students using their smartphones to capture the beauty of the old streets, migrant workers selling handcrafted goods from their home provinces, young professionals gathering in contemporary tea lounges, and elderly residents performing traditional face-changing opera for small audiences. Ciqikou was evidently more than just a conserved historical landmark; it was a microcosm of Chongqing, a city that welcomes change while clinging to the myriad tales that comprise its history. It was an area where the various migration waves had left their imprints, and those imprints were continually being reinterpreted.

I considered how this walk helped me better comprehend Chongqing's identity as we departed Ciqikou. Different histories and cultures cohabit, blend, and occasionally even vie for space in this metropolis where no one storyline is dominant. The diversity of Chongqing is beautiful because it defies easy categorization. It is dynamic, multi-layered, and always changing, much like the city itself.

In the following section of our trip, we continue to examine how this urban fusion manifests itself in various Chongqing neighbourhoods, following the ways that modernity, culture, and migration continue to influence the city's multi-identity character.

9.2 Streets of Convergence: Where Cultures Meet and Blend

The identity of Chongqing is not solely based on past migrations or unique cultural legacies; rather, it flourishes in the ordinary places where various influences meet, blend, and produce something new. If Ciqikou showed how the city's history was formed by migration, our subsequent tour of Nan'an District showed how that same multi-identity is still developing in modern urban life. Situated just across the Yangtze River from Jiefangbei's imposing skyline, Nan'an is one of the city's most vibrant neighbourhoods. Here, relics of ancient trade coexist with upscale homes, global business centers, and cultural venues that unite people from all walks of life. It is a region that has always been a site of change—where the city's identity is redefined by the meeting of the old and the new, the local and the global, and tradition and modernity.

The lively riverbank section of Nanbin Road, which provides one of the most breathtaking views of Chongqing's skyline, is where we started our stroll. The contrast between the past and present was evident as one strolled along the promenade—classy eateries offering international fare were situated next to modest food vendors offering traditional Sichuan street delicacies. We came onto a traditional lantern shop at one point, where the proprietor painstakingly created elaborate designs utilizing techniques that had been handed down through the years. The next we went into was a chic café with Sichuan pepper-infused coffee and European-style pastries, owned by a young businessman who had studied overseas.

Perhaps the most obvious example of Nan'an's multi-identity is seen in its marketplaces. We left the pristine waterfront and entered Tushan Market, one of the oldest business districts in the area, with its winding lanes. As sellers announced their daily deals, the air was filled with the aroma of fresh produce and spices. Tushan was an unadulterated depiction of Chongqing's urban fusion, in contrast to Ciqikou's carefully manicured historical charm. While migrant workers from all over China set up stalls selling ingredients from their home provinces, more recent foreign influences, such as Thai spices and Korean-style street food, had also found a home. A butcher who had been selling meat in the market for more than thirty years started up a discussion with me. Over time, his clientele had changed from local families to young professionals seeking organic cuts and international students searching for specialty meats for their own cultural dishes. His shop, a small but well-maintained space nestled between contemporary convenience stores, symbolized the district's transition. He told me, "*The city changes, but we find ways to keep what we know*", which appeared to capture the spirit of Nan'an.

We continued our walk down towards Danzishi Old Street, which was formerly a major trading center and is today a busy business area where people from many cultural backgrounds come together on a daily basis. The Chongqing International Business Hub, where we came across a diverse range of expatriates, entrepreneurs, and IT workers—all contributing their unique viewpoints and identities to the urban fabric—was what made this region so intriguing. We met a young Malaysian woman in a café who had travelled to Chongqing to launch a sustainable fashion company. She explained how she was able to work with local craftspeople and combine traditional Chinese materials with modern Southeast Asian designs because of the city's openness to creativity. She echoed the notion that the city's identity is not about uniformity but rather cohabitation when she stated, "*Chongqing doesn't erase differences—it makes room for them*".

At a nearby bookstore that specialized in bilingual literature, a group of international students from South Asia and Africa had assembled. The owner, who was born in Shanghai and has lived in Chongqing for a long time, has adapted to the changing demography of the city by collecting books and materials in many languages. I witnessed the actual meaning of variety in urban fusion in places like this, where people from all cultures could engage not only socially but also intellectually.

The scenery drastically changed as we went farther into Nan'an. Everywhere one looked was the juxtaposition between horizontal history and vertical ambition, with tall residential buildings rising over ground-level stores that had been there for decades. After passing an ancient mahjong house where senior citizens were chatting animatedly, we entered a co-working area where tech entrepreneurs were showcasing their most recent concepts. It was evident that Nan'an was both a district of cultural negotiation and economic prosperity. Here, people from many backgrounds found a space to coexist, where international influences were not just imported but also incorporated into local customs, and where traditional trades found new means to survive. The district was active because of the negotiation itself, even though the balance was not always perfect—some new developments drew criticism for upsetting local customs, while some older businesses battled against quick modernization.

I thought about how different this experience had been from Ciqikou as we finished our stroll. We had witnessed in Ciqikou how migration had left enduring cultural marks on the city and how the past influenced the present. We saw urban fusion in action in Nan'an, where tradition and change were dynamic processes occurring at the same time rather than distinct historical periods. The streets of convergence were conversations more than merely places. There, an international entrepreneur discovered methods to incorporate her roots into local business, a centuries-old butcher modified his craft for a contemporary customer, and a tech cluster coexisted with a market that sold produce from all across China. The identity of Chongqing was always changing in these settings where various individuals, viewpoints, and histories came together.

As I strolled through Nan'an, I realized that Chongqing's diversity was about letting the past and the future to coexist in the same area rather than about conserving either. The neighbourhood served as a microcosm of the city's capacity to reconcile inconsistencies, to build upon identities without erasing them, and to serve as a location where cultures coexist rather than clash.

I took with me the knowledge that diversity in Chongqing is not merely historical or demographic; rather, it is ingrained in the city's streets, shops, marketplaces, and daily discussions as we proceeded to the next phase of our adventure. It is a city that lives on diversity rather than merely accepting it.

9.3 Senses of Diversity: Language, Cuisine, and Social Interactions

Walking through the streets of Shapingba, I could feel the pulse of Chongqing's diversity in a different way. Here, in the university district and the surrounding migrant enclaves, the city's fusion was alive in the sounds of its languages, the scents of its kitchens, and the rhythms of its everyday conversations. If Nan'an had shown the meeting of old trades and new business, and Ciqikou had traced the historical layering of identity. In Chongqing, the sensory experience of diversity was not merely something to see; it was also something to taste, hear, and interact with personally.

The air itself was the first indication of this multiplicity. Fragments of several dialects drifted between groups of students sitting outside tea establishments and street food stalls as we moved through the university's streets. The forceful and expressive Chongqing dialect was still the most common, but it was laced with Sichuanese's milder tones, Mandarin with a Hubei accent, and the sporadic English or Korean utterances of the international student body. I paused to speak with a group of Xinjiang students, whose Mandarin reflected the unique northwest rhythm. They clarified that the city's welcoming attitude toward those in search of opportunity was just as much of a pull as the university's standing. One of them, a young lady majoring in business, told me that she had at first found the fast-paced local dialect of Chongqing difficult to understand but had eventually grown to value its expressiveness. "*You feel like you belong when you learn to speak like a local*", she remarked.

As we explored the nearby districts, where migrants from all around China have settled, that notion—of language as a gateway to belonging—became even more clear. Here, linguistic diversity was important for survival and adaptation as well as for scholarly exchange. While an elderly bookseller, originally from Guizhou, proudly noted that he had learned three regional languages since relocating to the city thirty years ago, a Hunan-style restaurant owner moved fluidly between her local tongue and Chongqinghua to receive orders from patrons. Each fresh wave of inhabitants introduced their own sounds into the city, changing the way people spoke in day-to-day interactions. This linguistic blending was not about erasure but about expansion.

From language to cuisine, the same patterns of fusion played out in the flavours that defined the district. Shapingba, with its student population and long-established migrant communities, had become one of the most diverse food landscapes in Chongqing. A Xinjiang-style barbecue station filled the air with the aroma of cumin and spice as we passed skewers frying over charcoal. To contrast with the numbing spiciness of Chongqing's well-known hotpot, a little Yunnan noodle store a few steps away promoted its specialty: rice noodles in a gentle broth. We found a small restaurant farther down the street, owned by a family from northeast China, whose dumplings were cooked to perfection in a cast-iron skillet. Food served as more than just nourishment at each of these locations; it was a declaration of identity and a means of establishing a feeling of home in a city that was always changing around its residents.

We ordered bowls of Shaanxi-style biangbiang noodles, their thick strands bathed in a fragrant, vinegary sauce, and sat down at a small outside table. The proprietor, a guy in his sixties, informed us that despite having relocated to Chongqing many years prior, he continued to create the local flavours. He pointed to a nearby eatery that served Sichuan-style stir-fry and grinned. He remarked, "*We all share customers*". They want spice on some days and something else on others. We can have both in the city. His remarks confirmed what I had sensed throughout our walk: that diversity in Chongqing was about embracing the various noises, flavours, and cultures that constituted everyday life rather than about separating people.

This mingling of cultures was evident in even the slightest interactions. A young Burmese student was assisted in choosing the appropriate tea leaves by a Jiangxi-born shopkeeper, who also explained the subtle variations among regional types. A Turkish seller of handcrafted jewellery engaged a local consumer in a lively discussion in Mandarin and English at a busy market, laughing while they haggled over a price. Meanwhile, a fruit vendor from Chongqing had learned enough Arabic to converse with a group of Middle Eastern students who came to his stand. These brief but profoundly significant events showed a city that promoted adaptation and interaction rather than absorption (See Figs. 9.8 and 9.9).

As the evening lights flickered on, we made our way toward a small food alley near the edge of the district, where university students gathered after long study sessions and night-shift workers stopped for a quick meal before heading back. I witnessed the genuine spirit of Chongqing's diversity here, at the nexus of so many diverse lives. A symphony of unstructured harmony was created by the sounds of discussion and laughter, which blended languages and accents. With each dish serving as a memento of someone's journey, the smells of grilled meats, spicy broths, and freshly baked buns formed an unseen map of migration stories. There was also no single tale in the

Fig. 9.8 It is true that through many walks and social experiences in the city, you will see the city has promoted adaptation and interaction more than absorption; something that you can only experience when you walk through communities, street markets, residential and commercial zones, etc. (*Source* Photos are taken by the Author in August 2024)

Fig. 9.9 A narrow alleyway is simply a place for a small street market for vegetables. This captures a typical urban market setting in residential communities, hidden from all those busy life of the touristic zones. (*Source* Photos are taken by the Author in August 2024)

common areas of this urban fusion, only the ongoing blending of various identities, each of which found a role in the city's developing narrative.

As I made my way back through Shapingba's streets, I came to the conclusion that Chongqing was special not just because it could accommodate diversity but also because it could make it feel at home. In this metropolis, social interactions—no matter how fleeting—were moments of cross-cultural connection, food was not alien but just another aspect of everyday life, and linguistic transitions were not interruptions but extensions of conversation. Ultimately, diversity in this place was not a show or an abstract idea; rather, it was a lived reality that permeated the streets, marketplaces, eateries, and discussions that defined the beat of the city.

9.4 Evolving Identities: New Generations, Digital Spaces, and Future Belonging

The sleek towers of Chongqing's contemporary cityscape gradually rose in front of us as we travelled toward Jiangbeizui, their acute angles reflecting the city's shifting landscape [7]. Although the past and present coexist in this district of Chongqing, the younger, digitally native generation seems to be holding the future in its hands.

It felt like the personification of the city's future. I could sense the city's changing pulse here, surrounded by soaring skyscrapers and glittering shopping centers, as it welcomed both new developments and new identities. This reality was influenced not only by the city's physical changes but also by cultural and digital influences that had altered the younger generation's perception of its position in the city and sense of belonging.

It was immediately noticeable how different Jiangbeizui's streets were from Chongqing's more historic neighborhoods. Here, the shops were more cosmopolitan, the buildings taller, and the streets wider. In the distance, the familiar features of Chongqing, the congested street markets, the interminable staircases, and the tight alleyways, were still there, but they seemed like remnants of another time. This area's landscape appeared to be characterized by a new rhythm, one that was fueled by a younger generation that was equally adept at navigating the digital and physical worlds. In many ways, the city was reinventing itself in real-time. I observed groups of young professionals walking along the sidewalks, talking on their phones as they made their way to slick cafes or the tall office buildings that appeared to rise like monsters made of steel and glass out of the pavement. Here, the modern and the old coexisted in a continuous dialogue. I heard bits and pieces of discussion as I passed, some in Chongqinghua, some in Mandarin, and some in English, all of which blended together seamlessly. What most surprised me, though, was how insignificant those linguistic distinctions appeared to be. The shared experience of the location and the shared engagement with digital platforms—which made it possible for these voices to converge in ways that were impossible only a few decades ago—were what counted.

Through my observations, I could see how the younger generation's habits reflected the way the digital age had changed how people engaged with their city. This is probably the same for everywhere else, but it felt quite different. It was impossible to overlook Jiangbeizui's urban environment, which appeared to have evolved into an extension of the virtual world. While virtual reality stores provided a preview of a future in which actual and virtual environments coexisted, digital billboards displayed ads in several languages. Outside of a contemporary retail center, I walked by a small group of youngsters seated at a table. They were all engrossed in their phones, browsing through apps that allowed them to stay in touch with the outside world. However, these exchanges had a sense of community despite the digital distance. The city had evolved into a place where the actual and virtual worlds coexisted, and the younger generation was the one that walked between both with ease, bringing their experiences of many identities with them.

There was a strong feeling of hybrid culture in these areas. I observed how young people maintained their ties to the distinctively Chongqing sensibility of warmth and family while embracing worldwide trends, whether they be in fashion, music, or technology. A young man carrying a takeout cup of herbal tea from the area emerged from a café sporting a sweatshirt with an English motto. In Jiangbeizui, a city now characterized by an increasingly fluid and mixed identity, the juxtaposition of local origins and global tendencies was evident everywhere. This was a site where new definitions of belonging were being created in addition to where modernity was emerging. As we moved through the area, I was struck by the presence of international businesses that had found their place alongside local enterprises. A cursory look across the streets showed tea cafes and hotpot restaurants selling Chongqing's famous foods next to Western fast-food businesses. This blending of indigenous and international flavors was not just seen in the cuisine; it was also visible in the architecture, the shops, and even the way young professionals lived and worked. In a city that had traditionally taken pride in its heritage [2], this hybrid culture was opening up new avenues for a sense of belonging. As a new generation envisioned and lived in a more interconnected, globalized world, the traditional ideas of what it meant to be Chongqingnese were being reinvented here in Jiangbeizui.

Here, identity development involved more than just adopting other cultures and new technologies; it also involved establishing environments that allowed for the emergence of new forms of belonging. At a pop-up art gallery, I started talking to a young woman who told me she had just returned to Chongqing from studying overseas. She claimed that she belonged to an expanding number of people who considered their identity as fluid and influenced by both their local and global experiences rather than being defined by a particular location or culture. As a digital marketing professional, she viewed the city's increasing interconnection as a thrilling chance to support Chongqing's ongoing growth. Her words stayed with me as I wandered through the more residential parts of Jiangbeizui, where towering apartment buildings were slowly replacing older neighborhoods. It appeared that the younger generation was adopting high-rise living as a means of securing space in a metropolis that was growing quickly in both size and complexity, rather than as a luxury. With their cutting-edge facilities and intelligent technologies, these new areas gave people who wanted to create new types of lives here a sense of potential. In many respects, these skyscrapers symbolized Chongqing's future as a city that was fostering its strong, long-standing relationships while simultaneously learning to integrate into international networks.

I couldn't help but feel positive and hopeful about the future as the evening went on. I could see a city changing into one that was more about the future than just its previous traditions in Jiangbeizui, where the ancient and the modern met with such tangible vitality. The younger generation, digital spaces, and new urban developments were shaping Chongqing's multi-identity character, which was no longer solely the result of its historical migrations or cultural fusion. These developments allowed the city to embrace its global potential while maintaining a strong connection to its roots [3], [6]. Last, but not least, Chongqing's identity fluidity was what most impressed me. People's experiences and expressions of belonging changed along with the city's growth and development. And I saw Chongqing's future develop in front of me as these changing identities—formed by new generations, new technology, and new ways of engaging with the city—took shape.

References

1. Bonato M (2022) The problem space of nature in Chongqing: a conjunctural analysis of environmental crisis discourses and local housing practices. Lingue Culture Mediazioni 9(1):165–188
2. Chen J, Judd B (2021) Relationality and territoriality: Rethinking policy circulation of industrial heritage reuse in Chongqing China. Int J Herit Stud 27(1):16–38
3. Cheng JY (2013) The "Chongqing Model": what it means to China today. Journal of Comparative Asian Development 12(3):411–442
4. Cheshmehzangi A (2020). Identity of cities and city of identities: Examples. In: Identity of cities and city of identities. Springer, Singapore, pp 179–244
5. Cheshmehzangi A, Li HMA (2020) Innovation through urban diversity and achieving comprehensive sustainable urbanism from a community-oriented approach. Curr Urban Stud 8(02):222
6. Cheshmehzangi A Tang T (2024) Chongqing: Navigating Rapid Urbanization and Regional Development Challenges. In China Under Construction: Shaping Cities Through Recent Urban Transformation, Springer, Singapore, pp 81–106
7. Jiang Z, Zhang S (2023) Research on international communication of chongqing city image in social media. Media and Communication Research 4:29–36
8. Wyman J (1997) The ambiguities of Chinese antiforeignism: Chongqing, 1870–1900. Late Imperial China 18(2):86–122

Chapter 10
Relations in Chongqing: Interactions and Connections Within a Multi-identity City

10.1 Cultural Interactions: Understanding the Dynamics of Diversity in Chongqing

As I walked through the bustling streets of Chongqing, it became clear that the city is more than just a collection of structures and roads; rather, it is a live, breathing example of the intricate and varied cultural exchanges that have shaped its identity. This dynamic fusion—a blending of traditions, cultures, and histories coming together in subtle and stunning ways—was revealed on every corner and street I examined. Because of its people and their relationships across generations, cultures, and socioeconomic spheres, Chongqing's multi-identity character is dynamic and ever-changing.

One cannot traverse the city without experiencing the influence of its varied communities, each of which contributes to the metropolitan environment [1–4]. This variety is a complicated web of relationships where many cultural identities not only overlap but frequently entwine, rather than just being a collection of discrete groups living together in one location. As we strolled through the city, I couldn't help but observe the silent discussions taking place at food markets, on street corners, and along the river, where people from many backgrounds congregate. Even if they are occasionally brief, these exchanges show the fundamental bonds that unite the many cultures and customs in this vast city.

The way that Chongqing aggressively celebrates its variety is among its most remarkable features. I noticed relics of the past when I strolled through the older, more traditional communities, including those in Ciqikou's twisting streets, where modern, cosmopolitan features coexist with traditional Bai ethnic architecture. The city's readiness to both assimilate and adjust to new influences while also preserving its historical identity is reflected in this blending of the old and the new, native and foreign. The city is simultaneously rooted in its past and looking to the future, as evidenced by its architecture, cuisine, and everyday rhythms. It is a place where the

A. Cheshmehzangi, *City Walks in Chongqing*, Urban Sustainability,
https://doi.org/10.1007/978-981-96-6042-1_10

Fig. 10.1 Ciqikou's twisting streets is more than just a marketplace or a regenerated old town. There is a strong sense of relations in this neighbourhood beyond just a regenerated project. *Source* Photo is taken by the Author in October 2019

old ways or traditions are not discarded but are rather allowed to coexist and, in some cases, inform or influence the new (Fig. 10.1).

The diversity I experienced during our walks is not always harmonious or without tension. Both integration and segregation are evident in various areas of the city, particularly in areas where more recent immigrant communities have made their home. Similar to the busy marketplaces in Nan'an District, these places are brimming with contrasts: kiosks selling cosmopolitan street cuisine, such Korean and Thai snacks, coexist with street sellers serving Sichuan-style hotpot. It is a place where many cultural components come together, sometimes to collaborate and other times to compete. This ongoing space negotiation serves as a reminder that these cross-cultural exchanges are not always simple or smooth, but it also demonstrates how ingrained the concept of diversity is in the city's consciousness.

These relationships are significantly shaped by Chongqing's migratory history [6]. People from all across China and beyond have migrated to the city over the centuries, each bringing with them their own languages, traditions, and lifestyles. In addition to establishing a city with multiple identities, these migrations have promoted a vibrant atmosphere with ongoing cross-cultural interaction. I frequently stopped at street corners or near riverbank parks on our walks to observe the small yet impactful ways that language, cuisine, and customs are shared. A young northern migrant worker might be greeted with a smile and an invitation to join a group of older guys who were conversing in the local dialect. Even though their origins and accents differ, a Chongqing family may eat dinner with a group of visitors from other regions of China, and their conversation will flow naturally. The city's multi-identity story revolves around these small, commonplace moments of connection.

The way that Chongqing's various ethnic groups interact with one another through interaction and a feeling of place, however, was what really caught my attention throughout our city walks. For instance, the riverbank serves as a meeting place for a variety of people, including business professionals traveling through on their way to meetings, young families strolling along the shoreline, and elderly inhabitants taking in a quiet afternoon. Interactions that cut over cultural barriers are made possible by this common area, which fosters a sense of belonging among many groups. I could see how the Yangtze River, which has played a significant role in the history and growth of the city, continues to be a symbol of connectedness as we strolled alongside it. In addition to connecting the city with the rest of China, it unites individuals from many cultural backgrounds and walks of life.

Another location where cultural exchanges can place in intriguing ways is Chongqing's marketplaces. We strolled through the congested alleys of Jiefangbei, where the more subdued scents of imported items blend with the flavour of Sichuan spices. Local and international cultures converge at this marketplace, where customs from all around China and beyond are incorporated into the city's business environment. Here, I witnessed the conflict between innovation and preservation as younger businesspeople provided goods influenced by international trends, while elderly store owners proudly showcased their traditional goods. Despite coming from a variety of ethnic backgrounds, the street vendors had a common language of trade that went beyond the particulars of their individual cultures, or cultural identities. At the same time, customers from different regions of China and outside mingled with one another in the market, which was also a place of subtle cultural exchange. This was frequently done through a mutual love of regional cuisine or handicrafts. Yet, Jiefangbei is seen as a major business and commercial hub of the city (Fig. 10.2).

Cultural exchanges in Chongqing encompass not only the relationships between various communities but also their relationships with their physical surroundings. The city exudes a strong sense of pride and ownership, but it is also acknowledged that its diverse population continuously reshapes and shares this space. It became evident when strolling through these areas that the city's identity is a dynamic, living narrative influenced by the relationships and interactions among its residents rather than a static, monolithic entity. The rich cultural richness of Chongqing is a result of this ongoing conversation between the past and present, the local and the global, the old and the new.

Fig. 10.2 Till today, Jiefangbei is still seen as the most important node for businesses and commercial activities in the city. *Source* Photo is taken by the Author in October 2019

Fig. 10.3 There is essence of history and relationship to the nature in many physical locations of Chongqing. *Source* Photos are taken by the Author in September 2020

Finally, the city walks gave me the opportunity to see the complex and continuous negotiations that shape Chongqing's multi-identity narrative in addition to giving me a chance to experience the city's physical locations (see Figs. 10.3, 10.4 and 10.5). Whether it was over a dinner, a chat, or just some quiet time, the people I encountered along the route served as a reminder that diversity in Chongqing is not just a concept; it is something that is experienced on a daily basis in every part of the city. These instances of cross-cultural exchange are what distinguish Chongqing as a genuinely singular location—a city where identity blending is a dynamic, ever-evolving process rather than merely a historical reality.

10.2 Social Networks: The Role of Personal Connections in Shaping Urban Life

During our city walks, I started to observe something remarkable as I walked through the lively streets of Chongqing: the complex, frequently imperceptible network of connections that bind people to the natural and constructed surroundings. These

Fig. 10.4 The roots meet the ground in a place where edge and the street are well defines spaces for movement and urban life. *Source* Photos are taken by the Author in September 2020

Fig. 10.5 This shows nature plays a major part in this mountainous city. The scene captures the interaction between nature and urban environments in a rather unique setting. *Source* Photos are taken by the Author in September 2020

Fig. 10.6 The lively streets of Chongqing are places of interaction and socialisation. It is hard to miss statues of little pandas crawling up to trees and walls as well as some art installations around the city. *Source* Photos are taken by the Author in 2021

networks—formed by friendships, family, employment, and day-to-day living—intertwine with the physical areas they occupy in Chongqing. The natural and man-made landscapes of the city play an active role in the development of social bonds and identity rather than just serving as the scenery for everyday living. The profound, nearly personal connections between locals and their surroundings became evident to me as I observed how people interacted with their environments during these excursions (Figs. 10.6 and 10.7).

The interaction between people and the city's geography is most noticeable at the center of Chongqing's urban sprawl. In addition to being topographical features, the city's steep hills and meandering rivers are essential components of its social structure. As I strolled through neighbourhoods atop the mountains, I closely observed how people's relationships were influenced by their surroundings. The residents are bound to their city by a shared memory that includes the winding lanes, the terraced houses that cling to the mountainside, and the constant view of the Yangtze River below. Chongqing's streets are vertical, which has created a special sense of community (see Fig. 10.8). As people climb or descend the steep slopes, neighbours regularly strike up spontaneous talks. The process of physically traversing the city and negotiating its slopes appears to represent the ongoing work required to preserve social bonds in this place—a continuous negotiating of space that reflects the complexities of relationships in a culture with several identities.

Fig. 10.7 Panda, a symbolic animal of the neighbouring province (i.e., Sichuan), cannot be missed from art installations in major public places and touristic parts of the city. *Source* Photos are taken by the Author in 2021

In this intricate network of relationships, the river is also quite important. In addition to being geographical features, the Yangtze and Jialing Rivers serve as links between Chongqing's various neighbourhoods and communities [5]. I witnessed firsthand how the waterway serves as a gathering place for many groups, where the boundaries between social, professional, and personal lives are blurred, while strolling along the riverbanks. In the same area, families walk along the waterfront, residents go about their everyday lives, and visitors enjoy the scenery. In addition to being beautiful places, the riverbank regions play a crucial role in interpersonal interactions. The river provides a more contemporary and recreational function, a place to unwind and socialize, but for the younger generations it evokes memories of a time when it was the lifeblood of trade and transportation. All of these viewpoints are brought together by the river, which serves as a meeting point for many identities and experiences.

During another walking tour (or city walk) around Jiefangbei's busy sector, I was struck by how much the city's business environment influences interpersonal relationships. Here, people's relationships are frequently closely related to the built environment, which includes office buildings, malls, and high-rise structures. However, professional ties are not the only social networks that form in these kinds of settings. I saw how merchants, consumers, and even bystanders mingled in the congested

Fig. 10.8 The vertical streets and communities create a strong sense of relationship and closeness in residential areas that are from decades ago. *Source* Photo is taken by the Author in September 2020

shopping avenues and outdoor marketplaces, creating fleeting yet significant connections. These conversations frequently went beyond the actual purchase and turned into cordial discussions or suggestions regarding regional traditions and customs. These networks, which were created through routine interactions, are important to the social life of the city because they bring individuals from all walks of life together. Particularly in a business district like Jiefangbei, it became clear that the built environment supports a wide variety of relationships, both personal and professional, that support the city's changing character (see Fig. 10.9).

I also became more aware of how the urban planning and architecture of the older districts reflect the interactions between people and their surroundings as I strolled through residential areas. The building itself appeared to promote a deeper, more personal kind of social connection in areas with little alleyways and narrow roadways. People live, work, and socialize in close quarters, which creates

Fig. 10.9 There is a strong connection between the city's iconic buildings and features and its main zones and districts, such as Jiefangbei. *Source* Photo is taken by the Author in August 2024

a level of relational intimacy that is uncommon in more expansive, contemporary urban settings. Personal relationships—between neighbours, store owners, and onlookers—frequently develop on their own in these settings. These areas' communal character encourages a strong sense of community and belonging, which facilitates the more natural development of relationships. By assisting people in navigating their everyday lives, these personal networks—which are frequently focused on proximity—help create a feeling of community and shared identity.

But I was able to observe how social networks are always changing in the more recent, contemporary projects, such the high-rise residential buildings in the Yuzhong District. Residents of these structures, which are intended to house hundreds of families, frequently do not know one another when they initially move in. But eventually, links are formed by the common experiences of residing in these vertical groups. As I looked at these areas, it became evident that relationships there are frequently based on similar needs and experiences, such as using elevators together, interacting in lobbies, or exchanging ideas at neighbourhood convenience stores. Although these areas might not be as intimate as more established, conventional communities, they provide a foundation for the development of new social networks. But here, the difficulty lies in preserving a feeling of community in an environment where establishing personal ties might occasionally be more difficult. However, the rise of local social organizations, whether they are focused on children's activities, hobbies, or neighbourhood gatherings, demonstrates how individuals can adapt and find methods to build relationships in even the most impersonal settings.

The way the city's varied physical environments—its hills, rivers, historic neighbourhoods, and contemporary districts—cooperate with the social networks that develop inside them was what most impressed me throughout these walks. The city of Chongqing is based on the idea that locations are not neutral and that people's relationships to one another are influenced by the places they live in. As such, its urban architecture is fundamentally relational. These areas, which range from the narrow streets of historic districts to the slick, contemporary buildings in the heart of the city, encourage ongoing interaction between people, families, and communities. As we explored the city, it became evident that the natural and constructed settings actively influence the connections that characterize Chongqing's social life rather than only serving as passive backdrops. Because of their history and design, these areas serve as the cornerstone of a city where diversity and multiple identities are not only experienced but also ingrained in daily life. There is a reciprocal interaction between people and place in Chongqing. The social fabric of this multi-identity metropolis is created by these continuous exchanges [7], which are a discourse that is always changing as the city expands.

10.3 Urban Relations: How Shared Spaces Foster Social Connection and Unity

The significance of the city's common areas—parks, streets, and marketplaces—where the various strands of the urban fabric converge to tell a tale of social cohesion and solidarity became more and more apparent to me as I travelled farther into the center of Chongqing. These areas, which are frequently teeming with activity, serve as gathering places for individuals from all walks of life, including locals and visitors, elderly citizens and recent immigrants. They are more than just transitory spaces; they are dynamic, living spaces where bonds are forged, fostered, and occasionally even put to the test. As I moved through these areas, I saw how important they are in determining how individuals from various cultural backgrounds interact with one another and the city.

The parks along the Yangtze River were among the first locations I came across where these relationships were evident. I saw how these areas were utilized differently by several generations as I strolled around them. Chongqing's elderly population, which had been there for decades, congregated in groups to do tai chi, play mahjong, or converse. While street sellers sold regional snacks, tourists enjoyed the picturesque views while young families strolled with their kids in the same park. There was an amazing sense of community in these common areas, despite the glaring disparities in age, background, and reason for being there. It was not unusual to witness a local artist conversing with a foreign visitor or a granny conversing with a young couple. The park served as a neutral location for these disparate groups to congregate, engage, and build relationships, illustrating how communal settings can promote understanding and social integration.

I saw how the natural setting influenced these encounters as we strolled around the different parks. People were able to interact with one another through more formal activities or more informal chats because of the open spaces, the river, and the dense foliage. In these parks, humans and the environment have a relationship that goes beyond simple recreation; it is about establishing a common area where people from various ethnic backgrounds can feel included. In addition to serving as actual sites for recreation, these green spaces also serve as venues for social interaction and the ongoing formation of new identities and shared experiences. Because these venues encourage encounters that cut across generational and cultural divides, the way individuals use them reflects Chongqing's variety.

Chongqing's marketplaces provide yet another striking illustration of how communal areas promote interpersonal relationships. I observed how the bustling, lively spirit of the local markets, like those in the Yuzhong District, served as a melting pot for individuals from all walks of life as I meandered through their congested alleys. Every corner seemed to be alive with discussion, and the air was heavy with the aroma of fresh produce, spices, and sizzling street food. Customers spoke a variety of languages and dialects with local sellers, many of whom were members of ethnic minorities or immigrants. They created a space where cultural diversity was not only embraced but also celebrated by selling a variety of products from Sichuan specialties to items from other regions of China. I got a real sense of how the city's various communities interacted on a daily basis in these marketplaces, where individuals would bargain over prices, offer advice, or just chat about the weather.

These marketplaces are a great illustration of how public areas may serve as forums for social and economic interactions because of their outside stalls and unofficial interactions. The fact that the transactions in these markets frequently involved more than just business dealings really got my attention. These areas served as gathering places for people from all over China, from the countryside to the city center, where they engaged in deeper social interactions in addition to buying and selling commodities. I witnessed how a straightforward purchase may spark a discussion about hometowns, family, or the best method to prepare a certain cuisine. Even though these encounters were brief, they demonstrated the city's overall social cohesiveness, where individuals from different backgrounds were able to find common ground through their everyday activities.

The importance of Chongqing's public areas—its squares, roadways, and pedestrian zones—as venues for social interaction was further highlighted by a stroll through its busy streets. The city's diversity was most evident on the streets, which were always moving. There was a fresh face, a new sound, and a new sensation around every bend. These exchanges were made easier by the streets' actual layout and architecture. Narrow streets and small, neighbourhood stores in the city's older neighbourhoods enticed people to pause and converse, fostering intimacy in what may have otherwise been an impersonal urban setting. Wide boulevards and retail centers in the more contemporary neighbourhoods, such as Jiefangbei, made it easy for people from all backgrounds to come into contact with one another, whether in the busy commercial districts, office buildings, or coffee shops. Despite having differing scales and designs, these areas were both used as venues for fostering relationships.

During our city walks, it was particularly evident how these common areas served as neutral forums for social interactions in a city characterized by its diverse population. They made it possible for unofficial meetings, light conversation, and—above all—the development of connections that may not have otherwise occurred. The social fabric of these places was enhanced by the presence of individuals from all origins, each of whom contributed unique experiences and viewpoints. The relationships I saw went beyond simple greetings and courteous conversations; they were profound ties that bolstered a stronger sense of solidarity and belonging. The greater social story of Chongqing is reflected in these small, apparently inconsequential moments of interaction. The city's public areas serve as a platform for commonplace interactions that cut over cultural divides and promote harmony in the face of variety. People from various social classes, ethnic backgrounds, and backgrounds can meet and build relationships based on respect for one another and similar experiences in these common areas. As I strolled around Chongqing, it became evident that the city's strength was not only its physical infrastructure but also its public places' capacity to promote togetherness and social cohesion.

The more I explored and observed these spaces, the more I understood how people and the built environment are interdependent. People's interactions with one another are shaped by the physical areas of the city, and the city itself is shaped by these interactions. Public areas in Chongqing serve as more than just places to travel or relax; they are also locations where social bonds are forged and fostered. As I proceeded to explore these areas, I witnessed directly how Chongqing's common areas promote interpersonal relationships, bridging gaps between individuals from various backgrounds and fostering a feeling of cohesion in a city known for its diverse population.

10.4 Intergenerational Relations: Bridging the Past and Present in a Transforming City

Chongqing is a place of constant transformation, and with this shift comes a delicate balance between the past and present, the old and new. It became evident to us as we walked through both old and new areas that Chongqing's connection between the younger and older generations is greatly influenced by the city's development. This interaction is evident in the built environment itself, where tall skyscrapers coexist with old temples and traditional marketplaces bustle alongside upscale retail areas. The interactions between generations and their perceptions of this city of several identities are framed by these areas, both new and old.

During our city walks, one of the most remarkable things we noticed was how the older generation appeared to be very anchored in the city's history and customs, but they were also becoming more conscious of the need to change. In Jiefangbei's winding lanes, where historic stores continue to provide folk crafts and traditional herbal cures, older locals frequently told younger onlookers about the city's history. It was a mixture of common awareness that the city's culture was changing and nostalgia. The physical alterations in the city served as a sobering reminder of how much had been lost for many senior people, in addition to being a source of annoyance. However, there was also a willingness to accept change. A young employee who had recently opened a coffee shop in the district was greeted by an elderly man leaning on his cane during one of our walks. The elderly man gave them advise on the region, its history, and its shifting pace as they struck up a conversation. It was an exchange that reflected the city's culture of multi-identity and was full of respect for the past and present.

In contrast, the younger generations appear to be more accepting of the city's changing identity. The energy was noticeably different when strolling through the more recent neighbourhoods, including Jiangbeizui, where contemporary skyscrapers reach the sky. Younger folks felt comfortable in this dynamic, forward-thinking setting. They moved with assurance that demonstrated their flexibility and ease in a Chongqing that was changing quickly. The sharp difference between their relationships and those of the older generation really got to me. In addition to interacting with one another, young people also engaged with technology and the built environment in ways that were unfamiliar to their elders. They checked into places on social media, navigated the city using mobile applications, and interacted with a digital environment that felt very different from the elder generation's tangible, hand-built world.

Nevertheless, despite their divergent perspectives on the city, there was a noticeable sense of intergenerational bonding, especially in common areas. Young people were also present in the parks we visited, such as the lovely Nanshan Park, where the elderly congregated to socialise and share time through multiple activities. They were taking photos, walking their dogs, or just taking a minute to relax. It was implicitly understood that everyone shared these areas. The older generation, albeit being a little confused by the speed of modern life, nevertheless found significance in these areas as places to congregate and interact with the community, while the younger generation appeared to value the elderly's presence as stewards of the city's cultural memory.

I remember a particular moment in Wulong, on the outskirts of the city, when an older woman offered to show us around a local market. She explained the history of the market and how, prior to the new developments, it had been the center of the hamlet as she led us through the tangle of stalls. Her remarks were an invitation to comprehend how the past influenced the present, not merely a recitation of history. The group's younger members, who were more used to the city's more sophisticated areas, paid close attention. It served as a link between generations, accepting the new vitality that the younger generation brought to the city while simultaneously acknowledging the passing of some customs.

The way the city itself mirrored this intergenerational bond really got to me. With its steep hills and meandering alleyways, Chongqing's distinctive topography serves as a sort of metaphor for how its residents deal with the challenges of generational transition. The elder generations have always had a special bond with the city's physical geography because they grew up negotiating its challenging terrain. The city's hills and rivers are ingrained in their memory, making it a familiar location to them. However, younger people interact with the landscape in a different way because many of them grew up in an environment that is gradually becoming more urbanized. They may view the city with surprise and excitement, as a place of limitless opportunities, but they also have to deal with the difficulties of living in a metropolis that is changing so quickly.

The city's public transportation system, which has experienced significant expansion and modernization in recent years, was a prime example of these generational dynamics. The new metro lines, which mark a change in how people navigate the city, are both a marvel and a challenge for senior citizens. Many elderly individuals still find comfort in the rhythm of the city's more conventional transportation routes and prefer to walk the streets they know so well. The metro system, however, was welcomed by the youth as a representation of the city's development. However, I saw that there was a subdued regard for the elder generation even in the more modern metro stations. In an effort to bridge the generational divide and uphold the respect for seniors that is still fundamental to the city's cultural traditions, young people frequently volunteer their seats to older passengers.

Chongqing is an intriguing site to examine through the prism of intergenerational connections because of this interaction between the old and the new, between tradition and modernization. Although the city is always changing, there is an underlying consistency that binds all generations: a respect for the elders, an understanding of the history, and a sense of belonging. Both generations can learn from one other, exchange tales, and close the gap between a city that is changing quickly and the cultural legacy that still characterizes it in these common areas, whether they are parks, marketplaces, or streets.

In many respects, Chongqing's intergenerational relationships mirror the city's larger story of multiple identities. The younger generation is carrying the torch for the city's future, while the older generation is in possession of the keys to its past. They build a city that respects its past and its future by working together to negotiate the challenges of urbanization and find common ground in common areas. As I strolled throughout the city, I came to see that these intergenerational relationships are about more than just coexisting; they are about forming connections beyond time and creating a metropolis where various identities, both past and present, continue to converge.

References

1. Amin A, Graham S (1997) The ordinary city. Trans Inst Br Geogr 22(4):411–429

2. Cheshmehzangi A, Tang T (2024) Chongqing: navigating rapid urbanization and regional development challenges. In: China under construction: shaping cities through recent urban transformation. Springer, Singapore, pp 81–106
3. Dreier P, Mollenkopf J, Swanstrom T (2014) Place matters: metropolitics for the twenty-first century. Revised. University Press of Kansas, Lawrence, KS
4. Katz B, Bradley J (2013) The metropolitan revolution: How cities and metros are fixing our broken politics and fragile economy. Brookings Institution Press, Washington, DC
5. Li Y, Cheshmehzangi A, He BJ (2023). A framework of urban renewal for leisure space in neighborhood streets: a case study of Shazheng Street in Chongqing, China. In: International conference on urban climate, sustainability and urban design. Springer Nature Singapore, Singapore, pp 693–705
6. Trenti RW (2021) Internal migration in China: origins, development, and effects on the city of Chongqing. Available online from: https://unitesi.unive.it/handle/20.500.14247/12796
7. Zhu C (2020) A bigger picture: the making of artists in contemporary China. Doctoral dissertation, The Chinese University of Hong Kong (Hong Kong)

Part III
The Multi-Locality City

Chapter 11
Old and New in Layers of Time: The Interwoven Layers of Chongqing's Multi-locality

11.1 The Persistence of the Old: Historic Neighbourhoods in a Changing City

Within their streets, alleys, and riverbank nooks, Chongqing's historic districts bear the scars of generations who have witnessed the city's numerous changes. As I moved around these areas, I frequently sensed the depth of their pasts, hidden beneath contemporary facades, new extensions, and the changing skyline. Many of these old neighbourhoods still exist, nevertheless, despite the city's unrelenting urban growth—not as artifacts, but as vibrant, living components of Chongqing's multi-locality (Fig. 11.1 and 11.2).

The inconsistencies present in this interaction between the ancient and the new were made clear by one walk in Ciqikou. With its sloping alleys and Qing-era architecture [1], the historic town, which was once a crucial river port, nevertheless retains remnants of its past. However, neon-lit cafés and boutique hotels now serve a surge of domestic tourists alongside the tea houses and artisan stores that have been conserved. The perfume of contemporary espresso blended with the ancient shao kao as I strolled down its busy main street, a clear example of how these places change over time without completely losing their historic character. With a quiet resilience, the vendors—many of whom are from families that have been in this area for decades—modernize without losing sight of their history (Fig. 11.3).

However, in other neighbourhoods, the enduring presence of the past is a lived reality rather than merely a cultural façade (see Figs. 11.4 and 11.5). I recall spending an afternoon on Shancheng Lane, where senior citizens still congregated outside their century-old houses to play mahjong and exchange news. In a city otherwise characterized by its rapid redevelopment, the alley is a historic relic that has undergone multiple renovations while maintaining its essence. I could sense the strong ties that bind locals to their surroundings here. Shancheng Lane is authentic because it continues to serve as a site of everyday living rather than just a tourist attraction, in contrast to the idealized portrayal of history found in certain restored neighbourhoods.

A. Cheshmehzangi, *City Walks in Chongqing*, Urban Sustainability,
https://doi.org/10.1007/978-981-96-6042-1_11

Fig. 11.1 Hongya Cave Cultural Center is one of the main attraction points in the city where the old meets the new, and where multi-locality of the city is preserved, showcased, and experienced. *Source* Photos are taken by the Author in October 2019

I discovered another remarkable example in Huangjueping, a fishing village that has been transformed into an urban area further south along the Yangtze. The neighbourhood has a multifaceted identity that blends Chongqing's industrial heritage [2] with its burgeoning creative sector. It is well-known for its streets covered in murals and the Sichuan Fine Arts Institute. As I made my way down the winding alleys, I came across relics from the past, such rusting factory doors, the remains of workers' dormitories, and old tea cafes where senior citizens still spend their afternoons. However, this environment was dotted with independent bookshops, student-heavy cafés, and brand-new art studios. Here, the conversation between the old and the new felt most natural, with none erasing the other but rather coexisting in a way that was distinctively Chongqing.

This cohabitation space has not been made available to all historic areas [6]. Rapid urbanization has occasionally resulted in the displacement of old towns, as commercial centers and high-rises have replaced them. An example of this reality may be seen in the way Nan'an District has changed. Once lined with classic courtyard residences and small family-run businesses, a substantial portion of the region has undergone extensive reconstruction. As I strolled through its more recent areas, I noticed how different they felt from the city's older neighbourhoods. However, remnants of the past persisted, such as a lone noodle restaurant tucked amid clean, minimalist cafés

Fig. 11.2 Hongya Cave Cultural Center is an example of regeneration and commercialized zones in the city where a destination is born. *Source* Photos are taken by the Author in October 2019

and an ancient shrine tucked between contemporary office complexes. The old refuses to completely go, even in areas where it has been totally eclipsed.

In Chongqing's multi-locality, history is not only reflected in the physical buildings but also in the people who still live there and how they adjust to the shifting urban landscape. In certain instances, such as traditional family restaurants that have modernized their decor while maintaining their century-old recipes, this transformation is smooth. In some places, there is a tacit resistance, with older neighbourhood inhabitants choosing to stay put in spite of growing living expenses and the encroaching development.

My learning from these city walks that the old does not just disappear when the new comes around. In response to the forces of modernization, it changes, finds new forms, and, in certain cases, makes an even stronger statement. Chongqing's historic neighbourhoods' tenacity narrates a tale of identity preservation among change, even as the city grows faster. And as I keep exploring these areas, I am reminded that multi-locality is about negotiating the past and future, what is left and what is emerging, all within the same cityscape. It is not just about separate locations coexisting.

Fig. 11.3 Ciqikou old town is a unique example of interaction between the ancient and the new. Here is an example of an older and narrow stone alleyway with steep steps between old brick buildings conveying a rustic and historical atmosphere of this regenerated zone. *Source* Photo is taken by the Author in October 2019

11.2 The Rise of the New: Expanding Horizons and Emerging Urban Centers

Entering one of Chongqing's more recent neighbourhoods is like entering a another age, where the future is developing at an incredible rate. Not only are the differences between the old and the new obvious, but they are also profoundly felt. As we made our way through neighbourhoods that were rapidly changing, it became evident that these places do not erase the past; rather, they build upon it, covering it with new goals. These extending horizons—where modern metropolitan centers with state-of-the-art infrastructure meet towns whose traditional landscapes have just lately been altered by high-rise developments—may be the clearest indication of Chongqing's status as a multi-locality metropolis.

The Jiangbeizui financial district, a relatively recent addition to Chongqing's urban landscape, is one such notable example. The district's well designed grid of offices, opulent homes, and commercial centers spoke of a vision of global connectivity as we strolled across its skyline of tall glass facades. However, beneath this contemporary exterior, remnants of the past can still be seen in how people use the area, not always in the form of preserved architecture. In order to bridge the seemingly sharp gap between the old and the modern, vendors continue to erect little breakfast stands

Fig. 11.4 Historical maps of the city are displayed in many parts of the city, open to public and visible to how the old is preserved and the new is introduced to the city. *Source* Photo is taken from the displayed wall in a public place, by the Author in September 2020

Fig. 11.5 The display of old and new iconic buildings, structures, and features of Chongqing in a shopping mall. *Source* Photo is taken by the Author in 2021

on the outskirts, construction workers congregate in shady spots to eat, and historic ferry lines that connect the neighbourhood to the city's historic center still run.

Yuelai, a new eco-city on the northern outskirts of Chongqing, similarly exhibits this change in spatial experiences. In contrast to earlier urban centers that formed haphazardly and naturally, Yuelai is a carefully thought-out sustainable living experiment. It seems very different from previous neighbourhoods where movement is

Fig. 11.6 The old often meets the new in an architectural form. *Source* Photo is taken by the Author in his first visit to Chongqing, in October 2019

defined by winding alleys and thick building clusters because of its incorporation of smart technologies, natural spaces, and pedestrian-friendly layouts. In contrast to the fast-paced, high-energy atmosphere of Jiefangbei or the busy activity of Shapingba, Yuelai is a place where people are meant to stop, engage, and experience the city in a more methodical manner (see Fig. 11.6). However, remnants of Chongqing's history can still be seen here. With little contemporary tea cafes reinterpreting classic aesthetics, the tea culture is still very much alive. The Yangtze River, which has shaped Chongqing's identity for centuries, continues to define the district's edges, reminding us that even the newest parts of the city are inevitably linked to their historical context.

Redeveloped areas also tell an evolving story of urban life. We witnessed how industrial history are being incorporated into new narratives of art, culture, and innovation [5] through the conversion of former factory districts into creative areas [3, 4], like Eling 2. With converted warehouses housing art galleries, start-up areas, and outdoor events, Eling 2 has transformed into a vibrant cultural center. Even though the historical buildings are still there, their purposes have changed to reflect the evolving goals of Chongqing's newer generations. The way that these locations continue to serve the city's residents in new yet recognizable ways is what makes this interaction between the old and the modern so important. The sound of live music fills the brick walls that once held the buzz of machinery, and the smell of paper and ink has given way to that of coffee and conversation.

The city's multi-locality endures in spite of these changes. These new urban centers add to the richness of Chongqing's geographical identity rather than homogenizing it. The city has risen in layers rather than in a straight line, with pockets of the new and the old coexisting and influencing one another. It is not uncommon to walk from a high-tech innovation park into a neighbourhood of classic wooden homes that still survive; it is just the way the city is changing. Wandering around the quickly evolving Nan'an District, where shopping centers and apartment buildings have altered the skyline, was one of the most eye-opening experiences. The social dynamics that the new developments created, in addition to the obvious difference between them and older residences, were what really got my attention. Older neighbourhoods continued to be centers of deeply ingrained communal life, even as new apartment buildings provided contemporary conveniences and lifestyles. Children played in the same pathways their parents used to use, people continued to congregate in open courtyards, and senior citizens sat in small groups conversing over tea. These villages continued to exist, adapting rather than vanishing as glass-walled condominiums grew up around them. It became evident that Chongqing's modernity does not take away from the core of its residents' daily rhythms. Rather, it gives the city's constantly developing story a new dimension and potential.

Lastly, these new urban hubs reshape Chongqing's definition of belonging. They coexist with the historical and cultural fabric that has long precast the city's identity while providing new chances for employment, recreation, and social interaction. Every step we took through these areas demonstrated not only the aspirations for Chongqing's urban future but also the tenacity of its history, which is still intertwined into the lives of individuals who live there and continue to influence it. The emergence of the new does not suggest that the old will vanish; rather, it indicates that the city is expanding in ways that are consistent with its spirit of diversity and multi-locality, constantly striking a balance between the past and the future.

11.3 Where Old Meets New: Urban Transitions and Layered Identities

The changing layers of Chongqing's urban fabric are depicted in this section, with particular attention paid to how new neighbourhoods define and shape the city's identity while preserving ties to its past. The ancient and the new are not strictly separated in Chongqing's urban landscape; rather, they intersect to produce something completely different. While exploring the city, we have come across areas where modernism and history coexist together rather than antagonistically. This is particularly noticeable in places that have reused historic architecture, transformed old streets into modern design studios, or turned industrial remnants into cultural treasures (see Fig. 11.7).

The adaptive reuse of Chongqing's former industrial zones serves as an illustration of this identity layering. Former factories and warehouses have been converted into creative centers that combine elements of the past with the vitality of the present. Instead of erasing history, the reuse of these sites protects and reshapes it, enabling the past to thrive in a metropolis that is rushing toward the future. Chongqing's changing urban identity is defined by the continuity created by the coexistence of recently built commercial high-rises with historic temples and traditional teahouses. The city's multi-locality is brought to life through these contrasts. Instead of being a contradiction, the conflict between the past and present is an essential component of the city's continuous transformation.

Fig. 11.7 Preserving the memory and cultural treasures of the past in a public building for people to access, take photos, and interact with all the displays. *Source* Photo is taken by the Author in December 2024

Another layer of change is seen as we go through our city's markets, which have been around for decades. Here, tech-driven companies that serve a younger, tech-savvy clientele coexist with street vendors selling traditional delicacies and craftsmen creating timeless designs. These locations continue to be important cultural pillars in the face of growing urbanization, serving as a link between ancient customs and contemporary goals. We can see that Chongqing is a city of constant adaptation rather than just contrasts by looking at the intersections of the ancient and the new. It incorporates its history into fresh perspectives on city life rather than discarding it. Chongqing's multi-locality is not just about distinct districts growing independently; it is also about how each locality maintains ties to the city's larger transformation while carrying its own distinct story.

As we walked through these spaces, it became evident that the city's identity is not static. It is shaped by the people who use it, the companies that reshape its streets, and the communities that decide what should change and what should be kept the same. Chongqing continuously reinterprets what it means to be both an ancient and a modern city, whether it is in the skyline where new skyscrapers and antique pagodas coexist or in a historic lane transformed into an arts area. Here, the old and the new collide—not as two distinct universes, but as entwined stories forming a metropolis that is constantly changing.

11.4 Beyond Preservation and Development: Rethinking Multi-locality for the Future

Walking around Chongqing reveals that the city is a dynamic interplay between preservation and evolution rather than merely a collection of historic neighbourhoods and contemporary construction. Chongqing's identity is shaped by a constant interplay between the stories woven into its historic streets and the aspirations mirrored in its tall skyscrapers. This last portion examines the future of this multi-locality, the conflicts between conservation and development, and how the city's various local identities will shape its next phase.

Chongqing's ability to embrace rapid change while clinging to its past is one of its defining characteristics. This coexistence is not conflict-free or smooth. Urban growth is putting pressure on many historic districts, leaving locals to choose between the promise of better housing, infrastructure, and economic prospects and their long-standing sense of community. One feels both resilient and inevitable when strolling through places like those close to the former ferry docks, where traditional wooden houses originally lined the riverbanks. While some homes have been replaced by construction projects that promise a new way of life, others still stand with their walls full with memories.

However, modernity in Chongqing does not necessarily imply erasure. History can sometimes be both preserved and repurposed, resulting in the creation of new multi-local identity nodes. The industrial belt's repurposed factories, which are now used as creative spaces, reinterpret their historical relevance without entirely erasing it. These locations' occupants—artists, businesspeople, and former industrial workers who now operate cafés—are essential to this change. Their existence guarantees that the city's legacy endures and instead finds fresh forms in modern life (see Fig. 11.8).

The tension between development and preservation is also evident in Chongqing's public spaces. Once vital hubs for social interaction, traditional marketplaces now coexist on the same streets as upscale business complexes (Fig. 11.9). However, the ancient continues to exist in unexpected ways in some parts of the city. I once came upon a mahjong house in an alley close to a recently constructed financial area, its inside resonating with the typical clatter of tiles and cigarette smoke. The players, who were primarily old, appeared indifferent to the glass-covered buildings that rose above them. The everyday habit remained unaltered despite the significant changes in the surrounding urban fabric, creating a remarkable contrast. These kinds of incidents serve as a reminder that multi-locality is about more than just places; it is also about the people who live there and their capacity to influence the city in both significant and minor ways.

How Chongqing handles these challenges will determine its destiny. The desire to become a global city and flourish economically will keep pushing the limits of urban growth, but the city's heart will always be connected to its deeply ingrained neighbourhoods. Finding methods to incorporate the new while preserving the importance of the old is the difficult part. Is it possible for rehabilitation projects to put social continuity ahead of aesthetics? Will the city's cultural institutions help to promote

Fig. 11.8 An interactive display of a preserved old building in a digital platform. *Source* Photo is taken by the Author in September 2020

intergenerational communication so that the next generation can comprehend and value the layers of history they live in? As we explore the changing environment of Chongqing, these are the questions that remain.

Adaptability is ultimately what multi-locality in Chongqing is all about. It is a city that is always changing. The relationship between development and preservation is a constant negotiation and rewriting of what it means to be a part of this place rather than a contest to be won. And as we move through its streets, both ancient and modern, past and future, we join that dialogue as participants, spectators, and storytellers in a city that defies easy categorization.

Fig. 11.9 A food court and marketplace that preserves the past in a new form of urban design within the commercial heart of the city. *Source* Photo is taken by the Author in December 2023

References

1. Chabrowski II (2022) Ruling the stage: social and cultural history of Opera in Sichuan from the Qing to the People's Republic of China, vol. 49. Brill, Leiden, The Netherlands
2. Cheshmehzangi A (2023) Regenerating local industries and urban experiences that matter. In: Mapping urban regeneration: city life experiences in Yunnan, China. Springer Nature Singapore, Singapore, pp 23–75
3. Li Q (2021) Study on the sustainable development model of industrial heritage. In: IOP conference series: earth and environmental science. IOP Publishing, vol 631, no 1, p 012033
4. Li Y, Cheshmehzangi A, He BJ (2023) A framework of urban renewal for leisure space in neighborhood streets: a case study of Shazheng Street in Chongqing, China. In: International conference on urban climate, sustainability and urban design, November. Springer Nature Singapore, Singapore, pp 693–705
5. Ye Y (2024) Abandoned industrial space in Chongqing, China: significance, meaning, and management. Doctoral dissertation, Queensland University of Technology
6. Zhou J (2018) The new urbanisation plan and permanent urban settlement of migrants in Chongqing, China. Popul Space Place 24(6):e2144

Chapter 12
Food Culture and Multi-Locality: Flavors of Tradition and Transformation in Chongqing

12.1 Roots of Flavour: Traditional Culinary Hubs and Their Lasting Presence

There is an obvious sense of continuity ingrained in the city's culinary culture as one navigates the winding lanes of Chongqing's old food streets. This fact is blended in the city's significant food tourism [1, 2]. A live tapestry of tradition is created by the sound of woks clattering, the rhythmic chopping of vegetables, and the distinct aroma of slow-cooked broth simmering in clay pots that date back a century. Chongqing's culinary culture endures because it is firmly anchored in the city's geography, history, and cultural memory, in contrast to other facets of urban change where the old frequently gives way to the new. The city's multi-locality finds one of its most powerful expressions in these crowded marketplaces and modest family-run restaurants.

The culinary customs of Ciqikou, a historic neighbourhood that has withstand waves of urban growth, are still there despite the uniformity that comes with fast industrialization (see Figs. 12.1–12.3). With menus unaffected by fads, the merchants here have been operating out of the same tiny storefronts for decades. A bowl of Chongqing xiaomian, made with the same attention to detail as generations before, evokes a feeling of place—a recollection of the city's steep hills, misty mornings, and sense of community. Eating here is more than just a transaction; it is an act of taking part in a greater story that, via something as basic as a bowl of noodles, unites the past and present, residents and tourists.

Chongqing's connection to its culinary heritage is about continuity in a constantly shifting environment, not just nostalgia. Consider the busy Jiefangbei food alleyways, where a blend of more recent arrivals and long-standing vendors demonstrates the city's changing tastes. The sour-spicy scent of suanlafen pierces the humid air as an elderly couple at one stall continues to make it with the accuracy they have perfected over the course of a lifetime. Younger cooks nearby are experimenting with different takes on hotpot broth, adding surprising tastes and gently stretching

A. Cheshmehzangi, *City Walks in Chongqing*, Urban Sustainability,
https://doi.org/10.1007/978-981-96-6042-1_12

Fig. 12.1 Food culture is the main and most visible part of Ciqikou old town (*Source* photo is taken by the Author in October 2019)

Fig. 12.2 Food production is part of everyday life in touristic and non-touristic zones. The city offers food in variety of tastes/flavours/ spices, colours, and small all the way around. (*Source* Photos are taken by the Author in October 2019)

the bounds of tradition without breaking its bonds [5]. History and innovation are always interacting in Chongqing's food scene, which is multi-local (see Fig. 12.4). This coexistence—the preservation of the old and the cautious introduction of the new [4]—is an example of this.

Markets continue to be essential to this cultural tenacity. Early mornings in Huangjueping's wet market show a routine that has not altered in decades: sellers call

Fig. 12.3 Food production and products are displayed at the street level and in thresholds between outdoor and indoor spaces of Ciqikou old town (*Source* Photos are taken by the Author in October 2019)

Fig. 12.4 Sometimes, I refer to cities like Chongqing and Changsha as large open-air restaurants. This is how I described the two cities just because of their very extended food or culinary culture and offerings, and many restaurants and places to eat and socialise based on food tourism and industry. The two cities truly offer the best dishes that one can find anywhere on the planet. I may be biased here but once must experience it to get to know the value of culinary culture in Chongqing. (*Source* photo is taken by the Author in October 2019)

out to devoted patrons by name, passing fresh food over wooden stalls while animated discussions take place. These markets are places of familiarity where generations of families have forged bonds over common culinary traditions, making them more than just places of business. These markets continue to exist in an age of supermarket convenience, not just because of habit but also because people recognize that food is more than just a source of food; it is a social tie that binds individuals to their community. However, the city's changing dynamics are visible even in these conventional centers. Interactions with food have been subtly altered by the emergence of digital food culture, the effects of delivery apps, and the growing influence of international eating habits. Although younger locals may not spend as much time in the markets as their grandparents did because to time constraints, the existence of these areas guarantees that their ties to Chongqing's culinary history will never entirely fade. The incomparable aromas of home-cooked meals that no internet purchase can match will continue to entice many people to return to these familiar locations throughout the holidays.

The ability of Chongqing's culinary culture to cut over class and geographic boundaries is what makes it so essential to its multi-local identity. Each area has historically evolved its own unique culinary subtleties due to the city's terrain, which includes steep inclines and scattered urban centers. Jiangbei's fusion of traditional flavours and contemporary inspirations, Nan'an's more subdued yet deeply ingrained family eateries, and Shapingba's thriving street-food culture all contribute to the city's overall narrative. Instead of causing fragmentation, the variety of these food hubs creates a network of interrelated culinary identities that make Chongqing as much about its flavours as its landmarks.

Exploring these food lanes and seeing how they provide as more than simply dining options is quite intimate. There is something deeply personal about the experience. A generational shift that occurs in many of these settings is exemplified by the old guy sitting outside his family's noodle store, observing as his son takes over the ladle. A closeness that transcends commerce is demonstrated when a seller meticulously modifies the degree of spice for a loyal customer. A shared cultural rhythm that endures despite the city's fast growth is revealed by the impromptu talks amongst strangers seated at a communal hotpot table (Fig. 12.5). These incidents, which occurred during our city walks, serve to further emphasize that Chongqing's culinary culture is about more than simply what is consumed; it is also about the connections and memories that are made over food.

Despite their firmly traditional origins, these gastronomic hotspots are not isolated from the city's larger development, which is perhaps the most startling discovery. They adapt to change and take in fresh influences, but they are still able to maintain their unique personalities. Chongqing's capacity to embrace modernity while staying rooted in its history is demonstrated by the tenacity of its culinary culture. The real spirit of multi-locality is most visibly experienced in these food streets, busy marketplaces, and historic restaurants—where the city's new and old identities come together, one meal at a time.

Fig. 12.5 Chongqing's hotpot can be found almost everywhere in the city. But it is truly unique to Chongqing, and second to none for sure. The busy nightlife and night activities of the city are mostly boosted by many 24/7 hotpot restaurants around the city. This photo is an exemplar, which was taken after 3am as if it is just 3 pm! (*Source* photo is taken by the Author in December 2023)

12.2 New Gastronomic Frontiers: Emerging Food Scenes and Culinary Innovation

This section examines how Chongqing's expanding districts, high-end dining spots, and fusion eateries are redefining the city's food culture, reflecting the interplay between globalization, migration, and evolving consumer tastes. The culinary scene of Chongqing is a constantly growing tapestry of new culinary frontiers rather than just a collection of traditional flavours and long-standing customs. The city has changed in recent years, and food culture is no longer limited to the tried-and-true routes of traditional restaurants and street vendors. Rather, it spreads into new neighbourhoods, fine dining venues, and experimental fusion kitchens that represent the vibrancy of a city with multiple localities by reinterpreting the familiar. As we traverse these evolving culinary landscapes, we come across the interaction of migration, globalization, and changing consumer preferences, all of which come together to subtly and boldly alter Chongqing's culinary character.

One of the most striking shifts in the city's food culture emerges from its newer districts, where modernity is not just an aesthetic but a way of life. For instance, the narrow lanes of ancient communities in Jiangbeizui contrast sharply with the financial hubs and gleaming glass towers. Boutique cafes and upscale dining establishments have established themselves here, serving a clientele that is becoming more and more international. These restaurants frequently combine regional products with foreign cooking methods to create cuisine that is uniquely Chongqing but has influences from throughout the world [7]. One such restaurant served me a hotpot-style risotto, which was instantly familiar yet reinterpreted in a Western culinary context due to its broth-rich, numbing spice. The chef, who is from Guangdong but had his training in Italy, talked about how he thought Chongqing was the ideal location for culinary experimentation because both locals and visitors were open to trying out fresh takes on the city's traditional flavours.

Beyond fine dining, street food itself has begun to evolve in response to new urban patterns and demographics. Hybrid dining halls have emerged as hubs in the quickly growing Yubei District, where a surge of students and IT workers has raised demand for high-quality, convenient cuisine. These areas offer a variety of cuisines outside the city's traditional mainstays by fusing traditional hawker-style food stalls with contemporary, minimalist decor and computerized ordering methods. A seller I watched make a Chongqing-style banmian with a surprising twist: it was tossed with truffle oil and topped with a soft-boiled egg. This addition illustrates how migration and exposure to different cuisines quietly alter the everyday culinary experience. These instances, which may be found in the areas between innovation and tradition, show how the city's culinary culture develops while holding onto its core values.

As various culinary traditions collide and blend into Chongqing's multifaceted and fluid character, migration has a significant impact on the city's gastronomic development. As I strolled through Nan'an's food markets, I saw sellers selling grilled seafood in the Southeast Asian manner next to shops selling lamb skewers in the Xinjiang style, all of which were accompanied by the ubiquitous Chongqing mala tang. Interesting stories emerged from conversations with sellers. One Yunnan-born guy described how he had modified his family's mushroom hotpot to accommodate the Chongqing's love of spice by adding Sichuan peppercorns and chili oil to satisfy the locals' need for numbing heat. The city's changing food culture is further enhanced by this blending of culinary identities, which is influenced by individuals moving between different areas.

Technology has also accelerated the rise of new gastronomic trends in ways that were unimaginable in the past. Innovative culinary ideas can now thrive without physical storefronts thanks to meal delivery systems, which have become ingrained in urban life. Cloud kitchens, which are virtual restaurants that only exist in digital space but process hundreds of orders every day, have produced some of the most creative meals I have ever had. These ghost kitchens serve specialized niche markets, such as those seeking fusion versions of traditional street food or plant-based Chongqing

Fig. 12.6 Food courts and street food are celebrated through multi-localities of the city as well as preserving the culture and the past (*Source* photos are taken by the Author in March 2024)

cuisine. In addition to changing how food and physical location are related, this change also highlights the city's multi-locality, since culinary culture is now dispersed across the urban fabric via digital networks rather than being restricted to a particular area (see Figs. 12.6 and 12.7).

There is still an intriguing conversation between the old and the new despite the emergence of these new dining environments. Instead of just abandoning tradition, many younger chefs and entrepreneurs are figuring out how to build upon and reinterpret it. In the revitalized Eling district, I remember sitting in a recently opened tea house where traditional Sichuan tea customs were combined with contemporary design elements—handmade ceramic cups, modern calligraphy on the walls, and a menu that served oolong tea with regionally inspired desserts with a French patisserie flair. These areas ensure that local culinary heritage is relevant in a constantly shifting urban landscape by embracing the past rather than rejecting it in a way that fits with modern urban lives.

Last but not least, Chongqing's rich culinary heritage is not erased by the rise of new gastronomic frontiers; rather, it enriches an already intricate and multifaceted food culture. The city's culinary industry is so captivating because it can accommodate both its long-standing customs and innovative, forward-thinking ideas [6]. I am continually reminded as I move through these areas—whether it be the busy hallways of food halls, the experimental kitchens of upscale restaurants, or the mobile

Fig. 12.7 Food courts are often decorated with historical and symbolic features to resemble the past and heritage in the most tangible way in the visual and physical settings of the city. (*Source* photos are taken by the Author in March 2024)

networks of cloud kitchens—that Chongqing's cuisine is not only a legacy but also actively influencing its future. This evolving culinary landscape, reflective of a city that is both grounded and boundary-pushing, embodies the essence of multi-locality, where old flavours and new frontiers coexist in a continuous state of reinvention.

12.3 Where Generations and Recipes Meet: The Social Fabric of Food Sharing

This section briefly investigates how food connects people across different localities in Chongqing, from family gatherings and communal hotpot culture to digital food trends and the role of social media in shaping urban dining habits. It was impossible to overlook the strong sense of camaraderie that arose from the shared tables, the bubbling pots of hotpot, and the age-old custom of eating as we strolled through the busy streets of Chongqing. Food culture in Chongqing has a depth that goes beyond basic nourishment because of the city's multi-locality nature, which includes a variety of communities and ethnic backgrounds. It turns into a gesture of camaraderie—a thread connecting neighbourhoods, generations, and even the city's always changing modernity. This chapter, which focuses on Chongqing's culinary culture, explains how food, in all of its forms, unites individuals from different parts of the city in ways that are both historical and contemporary.

Our city walks around the many areas of Chongqing were a potent reminder of the ways in which food serves as a social exchange tool, a bridge across cultures, and a common language. It became evident that food had always served as a unifying factor in the more traditional districts, especially those we visited in places like Shapingba and Yuzhong. Perhaps Chongqing's most famous dish, hotpot, is more than simply a meal; it is a social gathering and a ritual. This social dish easily crosses generations, whether it is prepared in a lively restaurant or a family home. As much about connecting as it is about the flavours is the communal experience of dipping meat and veggies into the simmering broth, where each person contributes their unique flavour (see Figs. 12.8 and 12.9). The long-standing hotpot parties we observed were firmly anchored in family customs, where recipes handed down through the ages had preserved the social mealtime ritual.

Newer culinary trends, particularly those influenced by social media, started to appear as we made our way through the streets close to the Chongqing Nanping neighbourhood. The city's social fabric was woven by more than just the typical hotpot. A new dining culture has emerged in Chongqing as a result of the growth of digital food trends, with platforms such as WeChat and Douyin fostering a culture where sharing culinary experiences is just as significant as actually eating. A visit to a

Fig. 12.8 Beef (and other) hotpots, street food, and food culture are simply the visible part of Chongqing. While Chongqing's food culture is widely recognised as the intangible heritage of the city, you are very likely to experience and observe more tangibility of the food culture than any other feature or characteristics of the city. Hence, the conception or idea of a "large open-air restaurant" stays very strong in my mind (*Source* photos are taken by the Author in 2024)

Fig. 12.9 An example of a regenerated building where hot-pot restaurants are key attraction points. The blend of flavour and smell can lead the path towards such hidden gems from far distance. (*Source* photos are taken by the Author in 2024)

chic, modern eatery in the Jiefangbei neighbourhood demonstrated how Chongqing cuisine has evolved to suit current palates. The way people eat is being changed by a new generation. Food is more than simply eating at these places; it is also about sharing, posting, and taking pictures, fostering online food communities. Food culture has become more accessible as a result of the link between eating and posting about one's culinary experiences on social media. It is no longer limited to the actual area around the table; rather, it has spread into online places where exchanging recipes, advice, and personal stories strengthens the idea that eating is a communal activity.

The coexistence of the younger and older generations within the cuisine culture was one of the most remarkable things we noticed throughout our stay in Chongqing. It was common to witness young and old participating in the same culinary rite alongside one another. On the one hand, the elder generation continues to be staunch in their adherence to tradition, developing and savouring recipes that have been passed down through the years. In the meantime, the younger generation embraces the quick effect of digital media and contemporary culinary trends to discover their own personality in food. For instance, I saw older customers tucking into steaming bowls of traditional rice noodles while younger folks used their phones to share pictures of their meals with friends during a stroll around the bustling Yuzhong District food markets. The generational gap within the larger issue of food as a social link was highlighted by the juxtaposition of these experiences. The younger generation leaned toward food's function as a symbol of belonging, a component of a larger online social network, but the elder generation was more interested in the food's inherent value—taste, texture, and family history. This merging of digital and physical, traditional and modern, is a reflection of the larger urban fusion process taking place in Chongqing, where food is a key component of the city's ever changing multifaceted identity.

We observed another aspect of food's ability to bring people together from diverse locations in more suburban settings, such as the lively communities surrounding the Nan'an District. Here, street vendors and local markets provide venues for unofficial gatherings, and food acts as a tool for neighbourhood integration. These areas, which are alive with the sounds of sizzling woks and the aroma of fried delicacies, are the natural settings for cross-cultural interaction. It was amazing to see how people from all over the city congregate in these culinary hubs and share meals that are a reflection of their cultural past as well as the local flavours and ingredients. People from a wide range of social backgrounds and even different Chinese regions were able to feel a sense of belonging in this energetic metropolis because to the communal environment of these areas. The cuisine, which combined Sichuan spices, Chongqing hotpot ingredients, and even Cantonese influences, reflected the city's multifaceted character. The shared meal of skewered meats and dumplings seemed to me to speak to an unspoken oneness that cuts beyond national boundaries, dialects, and cultural origins.

But probably the most impactful thing I took away from our city walks was seeing how food has changed to meet the opportunities and problems of the modern world, especially with regard to sustainability. It became increasingly clear that plant-based dining options were becoming more popular as we strolled through neighbourhoods like Jiangbeizui, where new buildings are continuously sprouting. The traditional culinary culture was gradually incorporating these more recent, ecological eating practices, highlighting how Chongqing is negotiating the demands of urbanization and environmental issues. For instance, plant-based hotpot options have grown in popularity, preserving the communal experience that characterizes Chongqing's hotpot history but also showing a growing concern in sustainable food and an openness to novel dining habits. It is evident that Chongqing's culinary scene is changing, fusing the city's long heritage of culinary culture with the contemporary need to be environmentally conscious.

The unquestionable connection between eating and identity in Chongqing was what most impressed me during our investigations. Food continues to be a reliable anchor in a city that is ever changing and where many generations and cultures converge. Through the warmth of a hotpot shared with family, the clink of chopsticks in a digital food post, or the sharing of recipes amongst neighbours from various parts of the city, it unites people across time and place and creates shared experiences that bond us together. Chongqing's cuisine genuinely captures the essence of the city: it is varied, vibrant, and firmly anchored in customs that have been practiced for many generations.

12.4 Beyond the Plate: Food, Place-Making, and the Future of Multi-Locality

Here, we provide a brief discussion on how food culture contributes to Chongqing's multi-locality narrative, considering the balance between preservation and change, and how culinary heritage will continue to evolve in the city's shifting urban fabric. As we ventured through the various districts of Chongqing, we reflected on how deeply food has embedded itself in the city's evolving identity [3]. Food is more than simply something people consume in Jiangbeizui's tall skyscrapers or Yuzhong's winding lanes,it is a major component of the city's dynamic urban fabric. Chongqing's food culture provides a distinctive perspective on the city's dynamic relationship with its multi-locality. Food customs are both maintained and changed as urban areas continue to change, influencing the social fabric, aiding in the city's place-making initiatives, and providing a glimpse into how the city may develop in the future.

As I traversed Chongqing's diverse landscapes—from residential complexes to expansive market streets—I was always struck by how potent a place-making tool food is as a cultural force. The city's culinary culture is in many respects inextricably linked to the communities where people live. The gastronomic pleasures that Chongqing's communities have to offer frequently define their uniqueness. Consider our walk through the Liangjiang New Area, where the modern and the ancient meet in almost lyrical ways. Due to the area's quick expansion, upscale eateries that appeal to a wider range of palates have emerged, providing a more cosmopolitan dining experience. However, the spirit of Chongqing's culinary legacy is subtly maintained even in these contemporary settings. These more recent restaurants use items from nearby markets to create modern takes on classic cuisines, such fiery hotpot broths. The city's larger balance between respecting its culinary traditions and adjusting to the needs of a quickly modernizing urban context was reflected in the experience of tasting these fusion dishes, which seemed like a moment of both continuity and change.

This delicate balancing act between innovation and preservation is not exclusive to the restaurant industry; it occurs in people's daily lives as well. I saw that, despite urban growth, food frequently serves as a means for various locales to preserve their own identities as we strolled around locations like the quiet residential communities of Beibei and the busy street markets of Nan'an. Older generations in these more affluent areas of the city hold fast to their traditional eating customs—recipes handed down through the ages, street vendors serving traditional fare like grilled pork skewers and Chongqing noodles. Certain meals exude a tangible feeling of pride, and in certain settings, food serves as a symbol of both individual and societal history. However, these traditional foodways are not isolated. These same areas have seen the emergence of new culinary trends in recent years. The most striking thing to me was how well the city's culinary identity has incorporated new food practices, especially those inspired by international trends like plant-based eating or modern dessert techniques. The core of what makes Chongqing cuisine so distinctive is being preserved as young people who grew up eating traditional Chongqing meals are now experimenting with new ingredients and cooking techniques.

When we visited areas like the Jiefangbei retail district, where commercial expansion has led to the emergence of food concepts intended to both preserve tradition and cater to a growingly diverse population, this junction of the old and the new became particularly evident. We observed how traditional cuisines, like Chongqing hotpot, were presented in new formats as we moved through the mall corridors, appealing to younger, tech-savvy customers. The introduction of digital food ordering platforms, which allowed users to order and personalize their hotpot using an application, demonstrated how Chongqing's culinary scene is adapting to technological advancements. However, there was also no denying the broth's rich flavour or the genuineness of its contents. A prime illustration of how food culture is developing in Chongqing, a city that is continuously navigating its multi-locality and pushing the limits of what it means to be a modern urban area, is this combination of digital innovation with culinary history.

The most intriguing aspect of our stay in Chongqing, particularly in neighbourhoods or communities like Shapingba, was how the city's food culture has evolved into a kind of social capital. Eating is a means of claiming one's identity and community. Food helps people make a name for themselves in a more competitive and congested urban setting, whether it is a young couple making a popular food film or an older passing down a recipe for spicy hotpot. The streets we traversed were more than just streets; they were gathering spots where people gathered, established communities, and forged relationships. Food serves as a unifying factor in these communities, a shared experience that enables residents to put aside their differences and discover points of agreement. This sense of community, which food promotes, has grown in importance as the city expands and becomes more diverse.

As I look to the future, I wonder how Chongqing's culinary culture will develop further in a city that is rapidly becoming more urbanized and attracting more foreign influences. The city's culinary culture will surely continue to be influenced by the younger generations, who are exposed to international culinary trends more and more through social media. Nonetheless, there is still a great desire to protect Chongqing's culinary culture. How can these various culinary traditions come together to form a future food culture that honours the past while welcoming the future as the city's urban fabric continues to change? As I walk around the different neighbourhoods of Chongqing, I felt hopeful about the transitions and the future of those older communities. This place exudes vitality—a feeling that food, as a cultural pillar, will keep evolving and adapting while being rooted in the city's past and sense of place.

The future of food culture in Chongqing, I believe, lies in its ability to harness its multi-locality and use food as a medium for not just cultural preservation but also for the creation of new forms of belonging. One thing is certain: Chongqing's food culture will continue to play a crucial role in its evolution, serving as a link between generations, communities, and the future, whether it is through the blending of flavours, the incorporation of technology, or the way food continues to influence urban experiences. The future of Chongqing, like its cuisine, will be a fusion of flavours, customs, and influences—a multi-locality that continues to respect its rich past while establishing new avenues for innovation—became evident as we strolled around the city, trying its cuisine, and seeing how it has changed over time.

References

1. Chen Q, Huang R (2015) Is food tourism important to Chongqing (China). J Vacat Mark 1(13):225–236
2. Chen Q, Huang R (2016) Understanding the importance of food tourism to Chongqing, China. J Vacat Market 22(1):42–54
3. Cheshmehzangi A (2020) Reflecting on urban identities. In: Identity of cities and city of identities. Springer, Singapore, pp 133–178
4. Cheshmehzangi A (2023) City life experiences: lessons learnt for future urban regeneration strategies. In: Mapping urban regeneration: city life experiences in Yunnan, China (pp. 239–246). Springer Nature Singapore, Singapore
5. Hu P, Wu T, Zhang F, Zhang Y, Lu L, Zeng H, Shi ZM, Sharma M, Xun L, Zhao Y (2017) Association between eating out and socio-demographic factors of university students in Chongqing, China. Int J Environ Res Public Health 14(11):1322
6. Ma J (2018, December) Landmark food and the construction of food tourism city of Chongqing. In 2018 3rd international conference on education, e-learning and management technology (EEMT 2018). Atlantis Press, Dordrecht, pp 292–296
7. Rock M, Mauro SED, Chen S, Iachetta M, Mabey A, McGill K, Zhao J (2016) Food production in Chongqing, China: opportunities and challenges. Middle States Geograp 49:55–62

Chapter 13
Nature and Water: The Lifeblood of Chongqing's Multi-Locality Identity

13.1 The Rivers that Define: Chongqing's Waterways as Urban Boundaries and Connections

This section examines how the Jialing and Yangtze rivers influence the city's topography, acting as links for trade, culture, and mobility as well as physical borders between districts and communities. It explores the significance of the rivers in connecting the city's various communities both historically and currently [6]. More than just bodies of water, Chongqing's rivers are the unseen designers of the city's multi-locality, forming its neighbourhoods, affecting its beats, and determining how people travel, congregate, and define their sense of identity. The intricate and flowing relationship between land and water, which has long shaped the spatial dynamics of the city, is seen when strolling along the riverbanks. Instead of merely splitting Chongqing into different areas, the Yangtze and Jialing Rivers create unique yet interconnected pockets of existence. The spirit of Chongqing's multi-locality finds its most concrete manifestations in these river-driven areas (see Figs. 13.1–13.4).

A complex story is revealed at every turn along the rivers. While contemporary bridges span the sea, bringing together areas that previously seemed like different planets, the city's mercantile past is whispered by the old wharves, some of which have been transformed into bustling waterfront districts. I remember seeing the currents weave into one another, as the history of many neighbourhoods do, as I stood at Chaotianmen, where the Yangtze and Jialing meet (Figs. 13.5 and 13.6) (also see: [4,7]. This intersection is more than just a geographical one,it represents the ongoing shaping of Chongqing's urban identity by the interaction of human activity and natural forces. The rivers' broad stretches serve as barriers that have historically divided towns, forming natural demarcations in some places. I recall noticing how, despite their close proximity, the districts on opposite riverbanks—one entrenched in ancient trading customs, the other a burgeoning center of contemporary development—had very different personalities. These distinctions were maintained by the rivers, enabling each location to develop at its own pace. However, the water also

A. Cheshmehzangi, *City Walks in Chongqing*, Urban Sustainability,
https://doi.org/10.1007/978-981-96-6042-1_13

Fig. 13.1 The spirit of Chongqing is vividly perceived and experienced in its rivers and river-driven areas (*Source* Photo is taken by the Author in September 2020)

Fig. 13.2 There is a spot to cross the city, there is a spot to walk, to sit, and to swim, all thanks to the rivers of the city (*Source* Photo is taken by the Author in 2021)

functions as a bridge on its own. The concept that Chongqing's neighbourhoods are not isolated is further supported by the fact that ferry journeys over the river are still in operation despite the abundance of bridges [1]. The rivers define, but they do not divide, they impose boundaries, but they also dissolve them.

Through our city walks, I was drawn to the various ways in which riverbanks have been repurposed over time. Traditional stilted homes still cling to the slopes in some locations, their foundations set into the Jialing's sloping banks. In others, recently constructed promenades beckon residents to congregate, work out, and observe the always changing skyline reflections dancing on the water's surface. At a teahouse on the edge of one of these promenades, I recall stopping to hear an old guy recall a time when these waters were more crowded with wooden cargo boats than the modern, sleek cruise ships. He recalled a period when the flow of the river determined the pace

Fig. 13.3 The riverside places are filled with history, heritage, and urban life. Often offered as walkable edges or riverside parks, these places are critical parts of the city. (*Source* Photos are taken by the Author in 2021 (left) and in September 2020 (right))

of life, when every community along the banks had its own marketplaces, docks, and way of life, with the river serving as its lifeblood.

The shifting nature of the riverfronts became glaringly obvious as we travelled from district to district. Modern high-rises border the riverfront close to Nan'an, their glass facades reflecting the water's light and reflecting a time of explosive urban growth. However, a short distance away, in more historic areas, stone stairways go down to the river's bank, serving as a reminder of the time when the city's main commercial areas were dictated by water access. The ways that various Chongqing neighbourhoods engage with their rivers illustrate the intricacy of multi-locality; some welcome the rivers as areas for leisure and recreation, while others cling to their industrial past, with abandoned warehouses and dockyards serving as silent reminders of a bygone era.

Additionally, the bridges that currently connect the city have a tale of their own. Each one symbolizes a distinct stage of Chongqing's urban development and has its own engineering and design. Crossing the Dongshuimen Bridge offers a panorama that spans centuries: a skyline that contrasts modern buildings with traditional rooftops, all against the Yangtze's constant current. Here, the river actively contributes to the city's change rather than merely serving as a backdrop. These bridges reshape local interactions, bringing people closer together while preserving each district's unique identity.

Fig. 13.4 The rivers act as veins of Chongqing; the livelihood, the urban life, and the localities (*Source* Photos are taken by the Author in 2021 (left) and in September 2020 (right))

Fig. 13.5 The node where the two rivers meet with the view of Chaotianmen bridge in far distance. This particular node has now become a major tourist attraction which is displayed well on social and digital media, too (*Source* Photo is taken by the Author in 2021)

As we strolled, we witnessed the city's pace has always been determined by the rivers. The city appears to change in reaction to the rising water levels during the rainy season, adjusting to the swollen currents as it has done for generations. Some places become more lively, as though the rising river gives them fresh vitality. On the other hand, the dry seasons disclose the riverbanks' hidden margins, exposing steps and walkways that are submerged for the majority of the year and providing a brief window into the past. These cyclical changes serve as a reminder that Chongqing's

Fig. 13.6 Where the two rivers meet in Chongqing is famous throughout the history. It is now even more famous for Chongqing Grant Theatre building as well as ferries and riverside activities (*Source* Photo is taken by the Author in 2021)

relationship with water is dynamic and always evolving, involving a conversation between history and reinvention, as well as between nature and urban life.

Observing how daily life develops along the rivers was one of the most eye-opening parts of our city treks. Water continues to play a significant role in local culture, as evidenced by the scenes of street sellers setting up shop along the embankments, fishermen casting their lines at daybreak, and elderly inhabitants practicing tai chi against the backdrop of slowly rolling barges. The city's longstanding image as a center of trade and business is further supported in certain neighbourhoods where the rivers are busy thoroughfares with ferries and cargo vessels slicing through the surface. In others, the water serves as a space for introspection and a slower pace in a city that is modernizing quickly.

These rivers are the connections that bind Chongqing's diverse identities together, making them more than just natural features. They give the city a feeling of continuity by allowing for its future while yet rooted in its past. The Yangtze and Jialing influence how Chongqing's neighbourhoods interact, change, and define themselves, whether as obstacles, bridges, or lifelines. As a result of our travels, I now view these waterways as dynamic forces that influence the city's multi-locality identity rather than merely as landscape features. The history of both past and present communities may be read in their currents, which reflect Chongqing's dynamic yet deeply ingrained character.

13.2 Waterways and the Shaping of Local Identity: From Riverbank Communities to Urban Spaces

Chongqing's rivers have always been more than just topographical features; they are the bonds that bind the city's various communities together. I frequently found myself tracking the lives that have influenced and been influenced by these waterways as I strolled along the riverbanks. In addition to dividing the city, the vastness of the Yangtze and Jialing Rivers creates the areas where new urban identities, groups, and customs are formed. The rivers continue to serve as both barriers and bridges, defining local character in ways that are as fluid as the waters themselves, from the old dock communities to the modern high-rises that now serve as markers of a new Chongqing.

There is still a rhythm to life in the older riverbank towns that is clearly connected to the water's rise and fall. For example, when I was strolling around Ciqikou, I was reminded of how early commerce was influenced by the water, since merchants used to transport products from boats to the busy market districts. The architecture and the way people congregate at the water's edge, their daily activities taking place against the changing tide, are still remnants of the legacy. Here, the enduring ties between the neighbourhood and the river—rather than nostalgia—keep the past real. The dynamics change as you move down. The relationship between water and local identity takes on a new form in Nan'an, where urban planning has reshaped the riverbanks. The sleek, contemporary waterfront projects show Chongqing's aspirational future. However, remnants of the city's long-standing relationship with its waterways may still be seen in these high-rise settings. The vibrant night markets that flank the promenade are a modern interpretation of the social vitality of earlier riverfront villages. This is where the past and present are most clearly visible, amid the neon reflections on the sea.

The development of the riverbanks in Chongqing is not consistent. Every locality and district has a unique narrative to tell about how the rivers have influenced its identity. Consider Tongyuanju, where historic piers have been converted into public areas, fostering a conversation between the past and modern urban life. I remember seeing senior citizens doing tai chi by the water in the morning; their slow, methodical movements reflected a sense of place that goes beyond the fast-paced changes of the city. Meanwhile, younger generations congregate in cafés along the river, interacting with the area in ways that are uniquely their own. A new form of cultural interaction, influenced by leisure, creativity, and a changing sense of belonging, is now made possible by the same river that was once a commercial route.

In Chongqing, a metropolis with several distinct locales, water serves as both a geographical feature and an identification symbol. We can view the continuities and contrasts that characterize the urban experience through the prism of the riverbanks. I saw how the riverbank in Jiangbei has been transformed into a well-planned social area, with pop-up events and weekend festivals for a temporary, international clientele. Here, tradition no longer defines the idea of local identity; rather, movement and the fluidity of fresh experiences built upon the past determine it. Water, however, continues to serve as a cultural anchor in some places despite these changes. The river continues to be a source of income in communities such as Shuangjiang. Here, fishing communities continue to function, but with the stresses of urbanization taking a greater and greater toll. As I stood by the docks, I saw boats coming back with their morning catch—a custom that, although it has faded, nevertheless has meaning for people who have never known anything else. Unlike the commercial districts, where the river is an aesthetic backdrop, here it remains an active force, dictating the rhythms of work and social life.

The intricacy of Chongqing's multi-locality is seen by contrasting these many interactions with water. Some neighbourhoods continue to cling to the river's historical significance as a source of connection and sustenance, while others welcome it as a site of contemporary reinvention. As I moved between these areas, I became aware of how water simultaneously divides and unites neighbourhoods while bringing people together for a common urban conversation. The sustainability of these changing identities is another issue brought up by the alteration of riverside areas. Although it creates economic opportunities, the growing commercialization of waterfront regions runs the risk of weakening the cultural bonds that once characterized them. Whereas unofficial dockside marketplaces used to flourish, polished boardwalks today serve a different demographic. Particularly in places where fast urbanization has altered the traditional relationship between the city and its water [3], the conflict between preservation and advancement is apparent.

As our city walks went on, it became evident that Chongqing's identity has been shaped by water in a continuous manner. In addition to bearing the burden of the past, the rivers also help the city develop by establishing new avenues for interaction and a sense of community. Whether in the vast metropolitan waterfronts of the contemporary metropolis or the little meeting places of historic riverbank settlements, the conversation between water and place is ever-changing and constantly redefining what it means to be a part of this multi-local city.

13.3 Nature's Urban Landscape as the City's Ecological Canvas: Green Spaces and the Flow of Life in Chongqing

The smooth transition between Chongqing's natural and urban environments is apparent when strolling throughout the city, as green spaces serve as living, breathing extensions of the rivers that have formed the city's identity in addition to being recreational sites [2]. Together with its rivers, the city's topography—which includes deep valleys and rolling hills—creates a dynamic biological system that both frames and supports urban life. Each district reinforces Chongqing's multi-locality identity by providing a unique interaction between green space and water.

One of the most striking examples of this intricate balance is the way parks and forested areas form an unspoken dialogue with the rivers. As we looked about, we noticed how these areas provide a crucial contrast to the intensity of high-density communities by softening the usually angular urban structure. Consider Eling Park, which is situated atop a hill and offers views of the Jialing and Yangtze River confluence. Here, people take a break from the bustle of the city and use the park as a peaceful haven where the rivers serve as a constant backdrop, bridging the gap between nature and humanity in a way that is distinctively Chongqing.

How nature is incorporated into daily urban life is further demonstrated by the addition of green corridors along the riverbanks. In addition to being pedestrian routes, Nanbin Road's pathways are community areas where locals congregate for morning tai chi, where street musicians perform in time with the river's rhythmic flow, and where merchants offer tea and fresh fruit. These areas demonstrate that green spaces are still essential to daily living even in one of China's biggest megacities, expressing a common appreciation for nature's presence. The way that various districts interact with these natural spaces—some as centers of culture, others as peaceful retreats—makes Chongqing's multi-locality clear. The city's ubiquitous rivers bind them all together.

We saw firsthand how antique townscapes blend in perfectly with the natural environment further upstream in locations like Ciqikou. A delicate balance between natural preservation and architectural heritage is maintained by the trees that border the historic streets, the tiny gardens nestled between buildings, and the peaceful courtyards concealed behind teahouses. In this context, nature serves as both an aesthetic element and a historical continuation, reflecting the ways in which water has influenced human settlements over the ages. Even while urbanization speeds up in other parts of the city, many neighbourhoods are able to preserve their unique identity thanks to the preservation of green spaces.

We observed how urban planners are consciously working to integrate green and blue infrastructure throughout our strolls through more contemporary neighbourhoods, making sure that water and nature are not neglected in the face of growth. At first sight, Jiangbeizui's clean skyline may give the impression that it is disconnected from the city's natural surroundings, yet upon closer inspection, patches of vegetation are thoughtfully incorporated into the city's architecture. River-facing plazas, tree-lined boulevards, and rooftop gardens provide brief moments of relaxation while quietly recognizing the value of nature in a changing urban environment. The thoughtful incorporation of green infrastructure into high-rise communities demonstrates the city's recognition that ecological identity should not be sacrificed for modernization.

It would be impossible to discuss Chongqing's green areas without bringing up Nanshan, the expansive mountainous region that provides a striking contrast to the city's crowded center. As we strolled along its meandering trails, we witnessed firsthand how this area functions as an ecological lung, purifying the city's air and offering a haven for biodiversity, in addition to being a place for enjoyment. While the contemporary cafes and boutique hotels tucked away in the terrain testify to a changing interaction between nature and modern urban lifestyles, the tea plantations that dot the hillside serve as a reminder of the city's agrarian heritage. Chongqing's identity as a city of many locations, where various relationships with the environment continue to define its destiny, is reinforced by these dualities: past and present, rural and urban, nature and development (see Figs. 13.7 and 13.8).

These ecological narratives are reinforced by the city's water supply, which connects green areas to the larger hydrological system that supports them. An urban ecosystem that lives on its interconnectedness is demonstrated by the way the Yangtze nourishes the verdant wetlands in communities like Yubei, the tiny streams that

Fig. 13.7 There is a strong link between many areas of the city and its mountainous and natural environments; something that visitors may know often see or experience in the first instance (*Source* Photo is taken by the Author in September 2020)

Fig. 13.8 Hills, rocks and nature are part of the city's old and new areas. Here is an example of the bottom side of Hongya Cave Cultural Center (*Source* Photo is taken by the Author in October 2019)

meander through suburban parks, and the stormwater management gardens that are sprouting up in recently constructed places. Intentional attempts are made to preserve this equilibrium even in the center of the city, where dense infrastructure predominates. For example, riverbank parks act as a buffer against urban sprawl, tree canopies reduce heat islands, and rain gardens reduce the risk of flooding. Green areas and water have a relationship that goes beyond simple coexistence; it is one of mutual dependency, influencing not only the physical surroundings but also the daily lives of those who live in Chongqing.

It became evident to us as we traversed these landscapes that Chongqing's natural areas are more than merely inconspicuous urban landscapes (Fig. 13.9). They have a significant impact on how various communities interact with their environment as active members of the city's multi-locality. Every area narrates a tale of adaptation, preservation, and evolution, whether it is found in the serene gardens of ancient districts, the bustling riverbank promenades, or the tall forests of the outskirts. Nature will continue to play a major role in forming Chongqing's identity for a very long time to come thanks to the way these green and blue threads combine to produce an urban form that is as flexible as the waters that define it.

Fig. 13.9 It is hard to imagine Chongqing's cityscape without its natural features, nature, greenery, and the rivers (*Source* Photo is taken by the Author in September 2020)

13.4 Adapting to Water: Chongqing's Evolving Relationship with Floods, Infrastructure, and Sustainability

Negotiation, adaptability, and survival have always been key components of Chongqing's relationship with water. Flooding has been both an inherent issue and a force that has created the city's resiliency because of the Yangtze and Jialing rivers, which define the city's configuration. From the tall concrete embankments in places where rapid urbanization has changed the environment to the traditional flood-prone wooden houses in older riverfront neighborhoods, it is evident how this relationship has changed over time as one walks along the banks in various districts. Every community along the water has a unique resilience memory; some cling to traditional flood adaption strategies, while others use technology to control increasing water levels.

The lower lanes in the old districts close to Ciqikou, where we have strolled along the riverbank streets, are dotted with signs that show the historical flood levels—quiet reminders of the natural cycles that have long governed city life. The recently constructed shoreline at Jiangbeizui, on the other hand, demonstrates a forward-thinking strategy. Green buffer zones and controlled embankments show how Chongqing is trying to live with water rather than fight it. Because of their histories, economies, and communal goals, these two disparate locations demonstrate how various communities negotiate their own relationships with the same rivers.

Further into the core of the city, our walks through Chaotianmen revealed how large-scale infrastructure projects aim to control the rivers that once dictated movement and settlement patterns. Although the flood barriers in this area act as a reminder of the city's vulnerability, they are also a component of a larger initiative to recover space for contemporary urban living. In addition to being a reaction to the effects of climate change, the waterfront's evolution shows how Chongqing's changing connection with water is intricately linked to its goals for development, tourism, and livability. The way a city like Chongqing navigates its two identities—one connected to the past, when people lived at the whim of the rivers, and one that looks to the future, where water is controlled, regulated, and incorporated into urban planning—was demonstrated in these areas.

Another dimension of adaptation may be seen in the smaller tributaries and natural drainage systems we have come across on our excursions, some of which are hidden behind residential clusters and others that flow through parks [5]. Even as the city grows around these smaller waterways, they act as ecological lifelines, preserving biodiversity. They also show another aspect of Chongqing's multi-locality: the micro-level interactions between various communities and nature, ranging from more recent planned projects in Yubei to ancient neighborhoods in Dadukou. While some of these waterways have been transformed into green corridors that combine ecological preservation and infrastructure, others have remained unaltered and continue to flow in the same manner for generations.

Chongqing's flood resilience is not just about physical infrastructure. The acceptance of seasonal flooding as a natural part of the city's rhythm demonstrates the equally significant significance that social and cultural adaptation play. Our city walks have shown us how businesses along the riverbanks work knowing that they might have to close for a few weeks each year, how locals remember stories of water encroaching on their homes as children, and how riverside parks and public areas are designed to recognize the inevitable rise in water levels. Chongqing's multi-locality is further enhanced by this awareness of nature's power, which is woven throughout the city. Different districts and communities have found their own ways to live with water—some embracing it, others controlling it, but all acknowledging its presence as an inescapable force.

Concerns about sustainability are also influencing Chongqing's strategy for water adaptation. Our strolls past recent ecological restoration initiatives, such the wetland parks being created along specific Yangtze segments, indicate that the city is starting to consider options beyond flood protection and toward a more harmonious connection with water. Restoring biodiversity, enhancing water quality, and creating space for urban environment are some of the goals of these regions, such as the recently refurbished Nan'an District. But these initiatives also raise issues of equality and access: who gains from these areas, and how do various communities strike a balance between urbanization and environmental sustainability?

The story of Chongqing's changing relationship with water is ultimately one of compromise. It is a conversation between the past and the present, between urban aspirations and natural forces, and between conventional adaptation and contemporary action. Because of the city's multiple localities, this conversation is never one-off. It is about engineering solutions in some locations and cultural acceptance in others. As we traversed these diverse landscapes, it became evident that although Chongqing's rivers and waterways continue to be potent natural forces, they also actively contribute to the city's identity, whether by influencing local culture, posing problems for urban planners, or acting as the constant background to day-to-day life.

References

1. Bao HX, Li L, Lizieri C (2019) City profile: Chongqing (1997–2017). Cities 94:161–171
2. Cheshmehzangi A (2020) Identity of cities and city of identities: Examples. In: Identity of cities and city of identities. Springer Singapore, pp 179–244
3. Cheshmehzangi A, Tang T (2024) Chongqing: navigating rapid urbanization and regional development challenges. In: China under construction: shaping cities through recent urban transformation. Springer, Singapore, pp 81–106
4. Han SS, Wang Y (2001) Chongqing. Cities 18(2):115–125
5. Lei X, Lu B (2021) Integrated water environment treatment: Mountainous sponge city and three gorges reservoir region. CRC Press, Boca Raton, FL
6. Lim KF, Horesh N (2017) The Chongqing vs. Guangdong developmental 'models' in post-Mao China: regional and historical perspectives on the dynamics of socioeconomic change. J Asia Pacific Econ 22(3), 372–395

7. Zhou W, Deng J (2014) Old Chongqing: city gate culture and its memories: city gate culture and its memories. East Asian Lit (in Korean:동아인문학), 28, 377–410

Chapter 14
People's Places: Walking Through Everyday Spaces and the Living Pulse of Chongqing

14.1 The Everyday Social Hubs of Chongqing: Examples of Side Streets and Hidden Corners

There is something deeply familiar about the way certain corners of Chongqing reveal themselves through daily routines. The city breathes with an informality that keeps its rhythm in the alleys between busy streets, the shaded areas beneath elevated freeways, and the surprising stillness of little courtyards tucked between high-rises. These are the locations that spontaneously arise, formed by habit and necessity; they are not famous landmarks or officially recognized public squares [4]. As I have been walking through them, I have noticed how these commonplace areas serve as social hubs that connect the city's various neighbourhoods and highlight Chongqing's multi-local identity (see Figs. 14.1 and 14.2).

One of my initial insights into the social geography of Chongqing came while strolling down a modest alley in Yuzhong District. An arm's length of narrow walkway opened into a collection of little tea shops and restaurants, each with low chairs placed in an enticing but disorganized fashion (Fig. 14.3). While playing cards in the middle of the afternoon, a group of old guys sat bent over a table, smoking cigarettes idly. At their side, a younger group leaned against a storefront window, deep in conversation, switching between accents as they navigated various social domains. It was a simple scene, yet one that revealed how these in-between spaces hold together the social fabric of the city. The city's multi-local identity stems from both its physical dispersion and the variety of relationships that are supported by these microenvironments.

The informal character of gathering places is considerably more noticeable outside of the core areas. I was intrigued to a temporary dining area in Nan'an that was sandwiched between a row of hardware stores run by families. It appeared to be just another pause in the line of stores, but closer inspection revealed that it was a well-functioning social center. A low wooden table, worn smooth from usage, was surrounded by plastic stools of various heights. While delivery drivers, temporarily

A. Cheshmehzangi, *City Walks in Chongqing*, Urban Sustainability,
https://doi.org/10.1007/978-981-96-6042-1_14

Fig. 14.1 People's places in Chongqing can be seen and experienced differently at different times. These are places where people interact and socialise with the most, and places of indoor and outdoor (urban) environments (*Source* Photos are taken by the Author in August 2024)

relieved of their rounds, leaned against the doorway in cozy solitude, the proprietors of nearby companies occasionally stopped by to share news and enjoy tea. It was implicitly understood that this was a common area that existed equidistantly between the public and private spheres, between business and recreation [2, 6]. Chongqing's urban life is characterized by this type of spatial adaptation, where residents utilize what is available and create spaces for relaxation and interaction that defy strict urban planning.

Naturally, the city's unofficial meeting places are greatly influenced by tea culture. In Chongqing, tea booths serve as venues for storytelling, negotiation, and intergenerational continuity in addition to being locations to have tea. I discovered an antique teahouse tucked away behind a row of tourist stores while out on a stroll close to Ciqikou. Inside, the city's pace seemed to slow down, and it was more peaceful than the main street, protected from the steady stream of tourists. Younger professionals seeking some peace and quiet were among the patrons, as were older males dressed in faded Mao suits. Here, the weight of familiarity was added to the steady pace of the exchanges. Some interactions were based on common histories, on tales of a Chongqing that was both past and present, while others were strictly commercial, such as talking business over a pot of jasmine tea. In an urban setting that is changing quickly, these teahouses—whether they are repurposed from traditional courtyard residences or nestled into old alleyways—serve as symbols of continuity.

Fig. 14.2 Here, an indoor environment of a medium-sized shopping center allows connectivity from the street to the underground spaces through similar style of outdoor environments. (*Source* Photos are taken by the Author in August 2024)

Fig. 14.3 Modern and traditional teashops as well as street food and restaurants are displayed so attractively and fashionably to ensure visitors and passersby residents/citizens have a chance to interact with a moment of pause (*Source* Photo is taken by the Author in 2022)

Roadside stalls that line the city's streets and turn pedestrian walkways into extensions of private residences and small enterprises are equally significant. These kiosks establish a dynamic link between migrant and permanent communities in places like Shapingba, where the university audience mixes with locals. I came across a seller selling freshly grilled skewers one evening while strolling down a side street close to a student housing complex. Students gathered in a relaxed line, conversing in between bites, and the air was heavy with the aroma of sizzling spices. A tiny wooden bench had been positioned directly behind the kiosk, providing an unplanned sitting space where strangers soon found themselves conversing. The older woman who had been operating the stall for more than ten years talked about how the area has changed, but not in a depressing way. Instead, she acknowledged that the city's vitality lies in its ongoing reinvention. These booths serve as extensions of the city's communal spaces, spanning the gap between permanence and flux, and are more than just locations for business.

In a little-known Jiangbei backstreet, I had one of my most memorable epiphanies. A group of women had met to make dumplings in a small square, which was connected by a natural corridor made up of a small arcade of stores. People stopped by, sitting on upturned crates to enjoy a meal, even though there was no sign indicating that this was a formal restaurant. The women laughed and greeted familiar onlookers as they worked in a smooth cadence. It served as a reminder that some of the most important social areas in Chongqing are not formally recognized as such. They develop via repetition, habit, and the straightforward act of creating a homely atmosphere.

Chongqing's multi-locality encompasses not just its size and the various districts that make up its terrain, but also the ways in which its residents designate their own communal areas. As I stroll around these commonplace social centers, I have grown to value the city not only for its striking skyline or its architectural wonders, but also for the way its residents claim and alter the areas they call home. The alleys, teahouses, and street-side seating areas are more than just functional locations; they are where the city's pulse is most palpable, where its true identity is negotiated and lived.

These locations are not isolated. The cadence of everyday existence and the imperceptible yet profoundly felt networks of connections that crisscross Chongqing's numerous locales bind them together. These unofficial areas will change as the city does, but their fundamental function—giving people a sense of community in a dynamic setting—will never change. And I am discovering new facets of the city's history that are concealed in plain sight with every stroll through the city and every time I pause to look, rest, or listen.

14.2 Markets, Malls, and Changing Rhythms: Commerce as a Social Connector

Walking through the marketplaces of Chongqing, I find myself immersed in the unmistakable rhythm of a city constantly shifting between past and future. Tucked hidden in little lanes or spanning wide courtyards, the ancient wet markets hum with a ferocity that does not seem to be affected by time. At the break of dawn, sellers set up their produce almost ritualistically, with baskets of chiles promising the city's distinctive heat, live fish flipping in shallow buckets, and wooden stalls full of veggies still shimmering with morning dew. A cacophony of voices fills the air, including those of stall owners announcing daily discounts, purchasers negotiating rates, and the occasional laugh between familiar faces. These markets are social hubs that are intricately woven into the identities of the areas in which they are located, in addition to being places of commerce. Every visit shows a complicated web of ties, from frequent customers who share stories as much as products to lifelong vendors who pass down their booths to the next generation.

Although they now coexist with tall shopping malls that have developed into their own form of social hubs, these wet markets remain popular as the city grows and changes. The malls serve as well planned areas of leisure and commercial spectacle, in sharp contrast to the wet markets, which flourish on the spontaneity of human connection. However, they are also essential to Chongqing's multi-locality in their own unique way. Social dynamics play out in unique yet strangely familiar ways inside enormous complexes like the high-rise shopping malls of Jiefangbei or the winding passageways of Paradise Walk. While families browse department stores stocked with the newest international brands, young people congregate in food courts, their chatter overflowing into the open atriums. These shopping centers are more than just places to buy and sell; they are miniature representations of modern Chongqing, where people's goals, way of life, and social customs are just as prominently exhibited as the products being sold.

Between the old and the new, there exists an intermediate space: the night bazaars and pop-up markets that punctuate the city's streets after dusk. The distinctions between tradition and modernity are fascinatingly blurred here. Under neon lights, vendors—some of whom also work in the wet markets during the day—reappear with a new assortment of wares at their stalls, including handcrafted crafts, street food, and desserts. Young businesspeople set up booths with vintage cameras, imported clothing, or used books to appeal to a different clientele that is lured to novelty and the vibrancy of urban nightlife. In these areas, I observe how commerce transcends simple economic activity and turns into a platform for cultural expression, allowing various communities to engage and contribute to the city's changing identity.

In my many visits to experience Chongqing's people's places, I got to see the many facets of many social and economic scenes [3]. In particular, this experience is vividly visible as I go through the bustling, dilapidated streets of a morning market, the chic interiors of a mall, and then the vibrant expanse of a night bazaar. A greater picture of how the city accommodates its multiple identities [1] may be drawn from the various ways people congregate, purchase, and sell in these areas. An earlier, more personal kind of business that is based on community acquaintance is preserved via the wet markets. The city's rapid economic expansion and changing consumer habits are reflected in the malls, which represent aspirational urbanism. Chongqing's urban change is characterized by a spirit of reinvention, which is embodied by the night bazaars, which are dynamic and always evolving.

Despite the differences in the landscapes and structures, there is a common sense of belonging in each of these locations. More than just actual places, the markets, shopping centers, and bazaars are expressions of the city's multi-locality [5], where the past and present coexist, where long-standing customs are modified to accommodate contemporary influences, and where trade serves as a social glue that unites people from all walks of life. My strolls around these areas show not just how individuals shop but also how they move through, engage with, and eventually influence the city in which they live.

14.3 Transit, Bridges, and Public Spaces: Moving Through a City of Many Centers

The city of Chongqing is dynamic. The intricacy of its transportation system, which not only provides movement but also defines its multi-locality, is equal to the complexity of its geography. Our city walks made it clear how transportation, whether it be by boat, monorail, or the meandering pedestrian bridges, influences how people live their daily lives. The urban fabric of Chongqing is dispersed over several hubs, each with its own distinct character, economy, and rhythm, in contrast to cities where a single central business district predominates. The way individuals move across these many zones reveals a tale of adaptability, practicality, and fortitude.

One of the most striking aspects of Chongqing's transit system is its reliance on the monorail. The rails go along cliffsides with expansive views of the river below in some locations, while in others they cut through apartment buildings. I was surprised by how well transportation blends into the existing environment as I stood on the platform of the Liziba station and watched the monorail vanish into the side of a residential tower. It is about movement as a collective urban experience, not just about traveling from one place to another. As they make their way through the everyday rhythm of the city, people board with baskets full of fruit, students go over their notes between stops, while office workers look at their phones. Beyond its usefulness, the monorail serves as a social equalizer by uniting disparate populations in a single moving area.

The river ferries that connect places that might otherwise seem worlds apart by crisscrossing the Jialing and Yangtze Rivers are equally significant. The ferry system has a slower, older pace that ties into the city's history as a water-centric trading hub, in contrast to the monorail's imposing modernity. I recall getting on a ship that seemed like a doorway to a different era during one of our strolls close to Chaotianmen. The pace was leisurely, and the significance of the water in the city's past was quietly appreciated. In addition to commuters, the ferries transport sellers, elderly people traveling across the river to see relatives, and visitors seeking an alternative view of the cityscape (see Figs. 14.4–14.6). The way different communities interact within these floating spaces reveals how water continues to function as both a boundary and a connector in Chongqing's urban fabric.

In this metropolis, bridges serve as more than just infrastructure; they are vital links that influence social and commercial relationships. Every crossing has an own narrative to tell, from the expansive, contemporary Dongshuimen Bridge to the more

Fig. 14.4 Chongqing's rivers are more than just the visible water bodies of the city. They are places where some locals come to enjoy, catch fish, and have a swim (*Source* Photo is taken by the Author in 2021)

Fig. 14.5 Chongqing's riverside and all its crossings are considered as people's places with excellent characteristics of localities. (*Source* Photo is taken by the Author in 2021

Fig. 14.6 The rivers have created many urban patterns and urban environments across the city. (*Source* Photo is taken by the Author in October 2019)

traditional, practical footbridges nestled amid residential areas. A number of these bridges have evolved into unofficial marketplaces, with traders erecting little stalls at their entrances, while others are simply stops for visitors to stop and enjoy the scenery. During one of our walks across the Shibanpo Bridge, I observed that the area served as more than just a means of transportation, but also as a sort of social space, with cyclists, pedestrians, and street vendors sharing the same concrete stretch. The bridge was not just about crossing the street; it was about staying, watching, and being in the ever-changing city.

This experience is further enhanced by public escalators and staircases, which are specific to Chongqing's mountainous environment. During one of our hikes, we rode the 112-m-long Huangguan Escalator, which is more than just a means of getting from one area to another. It serves as a symbol of how Chongqing's geography controls movement. Through these vertical channels, localities stay different but connected, businesses flourish in unexpected places, and residents adapt. Climbing or gliding down such escalators and steps allows one to take in the everyday choreography of urban life, as people pause at roadside food booths, weave through narrow alleyways, or change their itineraries based on weather and traffic.

The nature of a multi-locality metropolis is emphasized in every mode of transportation in Chongqing, from its swift monorail to the silent perseverance of its boats and stairways. The individuals who move through these areas, forming the character of their communities as they do so, are just as important as the infrastructure that links them. Chongqing is better understood as a city of mobility rather than just a city of places when one observes how the city's transportation areas serve as public areas where interactions organically occur and urban life is constantly negotiated.

14.4 Festivals, Gatherings, and Future Belonging: The Evolving Meaning of People's Places

The spaces of Chongqing are never static; they pulse with life, shifting in meaning and significance as the city itself evolves. If our city walks have taught us anything, it is that no place is the same, but despite this ongoing change, some customs and rhythms persist. The locations of Chongqing's inhabitants are much more than just actual places, whether it's early in the morning when tai chi practitioners congregate in the open plazas or during the Lantern Festival when the streets are illuminated by a cozy sense of shared memories. They are adaptive, transitory, and intricately linked to how citizens find a sense of place in a city that never fully establishes its identity.

The way festivals change ordinary settings was one of the most amazing things we saw. During important festivities, Ciqikou, which is normally a quiet place with slow-moving tourists and businesses, becomes a vibrant place with parades, performances, and ancient rituals. The smell of freshly steamed dumplings and crackling fireworks filled the air as I negotiated the crowds during Spring Festival. These social gatherings reinterpret public spaces, temporarily erasing the distinction between tradition and modern urban life. Through common traditions, the sidewalks where senior citizens typically play mahjong turn into festive thoroughfares that bring families and communities back together. Thus, this occurs in a set of shared rituals. It is in these moments that Chongqing's multi-locality is at its most visible—each district contributing its own unique take on the same celebration, each neighbourhood an individual thread in the fabric of a much larger urban story.

In addition to organized festivals, unplanned events in public areas also show how many communities contribute to the city's unique identity. Even though Jiangbeizui is dominated by contemporary skyscrapers, an evening stroll along the waterfront promenade shows a completely different side of the neighbourhood. Older couples walk hand in hand, groups of young dancers practice beneath neon lights, and groups of workers just finished their hours congregate around street musicians, allowing the music to serve as a little diversion from the daily grind. The area is claimed differently by each individual and group, but together, they help to define it as a gathering place for people from various social backgrounds and generations.

It is these organic, lived experiences that complicate the idea of fixed urban spaces. Consider People's Square's evolving terrain, for instance. Originally constructed as a large, formal municipal plaza, it has evolved into a flexible area for social protest, political expression, and relaxation. Its open space is used for group workouts by seniors in the morning, and by the afternoon, it has become a popular spot for professionals and students to congregate during breaks. The hitherto deserted hallways become bustling hubs of activity at night as food vendors occupy its perimeter. Every locale changes to suit its users, and no one description can ever adequately describe what a place is or will become. This flexibility allows locations to change in meaning and function over time, which is what makes Chongqing's urban identity so captivating.

New kinds of public spaces designed to function as future social hubs have also been added by urban development initiatives, especially in more recent neighbourhoods like Lijia. These planned urban areas were created with walkability and community engagement in mind, in contrast to previous districts that developed naturally. But as I stroll around these more recent areas of the city, I frequently ponder whether or not artificially created areas can accurately capture the essence of those that developed organically. Can an open square filled with art installations and curated greenery ever replace the raw vibrancy of a street corner where local vendors have been selling breakfast buns for decades? There can be more than one possible response. Rather, it appears that the urban fabric of Chongqing will always be a synthesis of emerging lifestyles and inherited customs (see examples in Figs. 14.7–14.9).

I was reminded by how very intimate these places are to the people who live there as we walked through Shancheng Lane's old lanes. In addition to being tangible relics of the past, the layers of history woven into each brick and the echoes of previous incarnations that continue to reverberate through the peaceful courtyards are also active contributors to the present and future of the city. At the same time, we have observed how contemporary public gathering places, such as the terraces of Raffles City and the pedestrian streets of Guanyinqiao, are creating new trends in public

Fig. 14.7 Art installations are part of urban spaces and create synergies between people and places. Some, as shown here, create a place to take photos and interact, and some are places to socialise (*Source* Photo is taken by the Author in 2024)

interaction. Places are no longer only inherited; instead, people create them and add new narratives to the already-existing urban fabric.

In this sense, the places of Chongqing are about time, relationship, and memory more than just geography. The city is a live, breathing thing that adapts to change while maintaining its distinctively Chongqing character. The fact that multi-locality is not only a structural reality of the city but also its defining pulse has been reaffirmed by this last tour of its numerous public areas, ranging from the oldest to the most modern. The social landscapes we have examined do more than simply reflect urban diversity; they also help to shape it, change with it, and guarantee that Chongqing will always belong to the people who use its streets, congregate in its squares, and find a home among its many people.

Fig. 14.8 With or without art, there is always a chance to interact with the public realm. Here, is an example of public space use for social media (*Source* Photo is taken by the Author in 2023)

Fig. 14.9 And sometimes, there are monumental buildings and iconic architecture that resembles people's places. A place where people could meet and catch-up after a year of not seeing each other; a place where they could sit and talk under the evergreen trees; and a place where people could interact and fall in love again and again. In every return, there is a meaning, and in each meaning, there is a sound of returning to the source. Some may see this in a moment, and for some, it may take a much longer time to realise. The many monuments of the city remain to create the urban fabric that promotes synthesis of emerging lifestyles and inherited customs. For me, this moment was truly a halt in time, like before, like many other times. (*Source* Photo is taken by the Author in his second visit to Chongqing, in September 2020)

References

1. Bell DA, De-Shalit A (2013) The spirit of cities: Why the identity of a city matters in a global age. Available online from: https://www.torrossa.com/gs/resourceProxy?an=5580054&publisher=FZO137
2. Carr S (1992) Public space. Cambridge University Press, Cambridge
3. Cheshmehzangi A (2014) Spatial syntagma and identity of a place: sensing, relating to, and knowing a place. J Hum Behav Soc Environ 24(7):799–810
4. Cheshmehzangi A, Tang T (2024) Chongqing: navigating rapid urbanization and regional development challenges. In: China under construction: shaping cities through recent urban transformation. Springer, Singapore, pp 81–106
5. Huang S (2018) Living in place: Rebuilding the relationship between people and place in Jiulongpo district, Chongqing, China (Doctoral dissertation)
6. Wynne D (1990) Leisure, lifestyle and the construction of social position. Leis Stud 9(1):21–34

Chapter 15
Chongqing: A Vibrant and Timeless Dragon that Never Sleeps

15.1 The Pulse of Chongqing: Energy that Never Fades

Chongqing is a city that never sleeps, where history and modernity, mountains and rivers, and tradition and innovation blend into a restless urban symphony. This book, *City Walks in Chongqing* (or to be known as *Chongqing City Walks*), has been a journey through the city's labyrinthine alleys, soaring bridges, neon-lit streets, and hidden corners. We structured all the work under three mappings of "multi-layers", "multi-identities", and "multi-localities", for which the City of Chongqing should be well regarded for, if this has not been the case yet. Through every walk over the many years of visits and mapping studies, we have uncovered the essence of a place that refuses to stand still—a vibrant, dynamic, and evolving metropolis that pulses with an energy unlike any other. My first journey to Chongqing goes back to October 2019 and the most recent was in December 2024. During this period, many city walks, mapping studies, and observation analysis have taken place. Hence, the book is the result of these five splendid years.

At its core, this book highlights Chongqing's multi-layered identity; a complexity beyond just a simple definition. Multiple migrations, changes in the economy, and unrelenting urbanization have all shaped this metropolis over centuries. Chongqing's history is one of reinvention, from its beginnings as a capital during the war to its current ascent as a leader in smart cities. We have looked at how the city's resilient and aspirational citizens define it just as much as its skyscrapers and waterways do.

The interaction of Chongqing's tradition and modernity has been a major issue in this work. Modern AI industries coexist with the city's historic streets and teahouses. Even as digital innovations influence its future, the smell of hotpot and the calls of street vendors evoke nostalgia. We have also looked at how Chongqing's identity is constantly being reshaped by migration and movement. Wealth redistribution [6], as it is now coined as *Chesh Theory* (with further explanation in sub-Sect. 2.3 of this chapter), has been a major factor in changing demographics, with long-time urbanites pursuing suburban lifestyles and former rural landowners becoming city

A. Cheshmehzangi, *City Walks in Chongqing*, Urban Sustainability,
https://doi.org/10.1007/978-981-96-6042-1_15

elites. Chongqing is an ideal case study for new urban theories because of this urban transformation, which questions accepted ideas about city growth.

Finally, the book is a tribute to the restless spirit of Chongqing, a city that defies categorization by embracing both order and disorder, the past and the future, permanence and change. Chongqing is truly a dragon that never sleeps, roaring through time and seasons with its vibrant night markets, mist-covered bridges, and changing cultural landscape. That essence is captured in this concluding chapter, which unifies the concepts we have studied to create a comprehensive portrait of a city that never dies. I recall myself thinking over and over about this beautiful city, articulating:

The inspirational assessments needed your overarching understanding, careful hypothetical evaluations, and trusted education. Desires on never-sleeping moments entangled in networks that harmonise inventiveness, sentimentalism and spiritualities, impacting nurturing communities in this majestic city. First, there are realities in quest, they have endless noteworthy treasures one bizarrely includes as superstitions, and many others that are yet to be uncovered in the once-in-a-lifetime connection not far away. Second, this is just a reality that ought to be faced with many moments and memories that genuinely matter. And third, Chongqing is the city beyond imagination that one must explore by walking through it with open eyes, a fresh mind, and a free soul, much more than what a book can offer under the title of *City Walks in Chongqing*. For that, I invite the readers to pay a visit to Chongqing and experience and observe what an urbanist once went through all its gems that it offers.

15.1.1 The City at Dawn and Dusk

The time of day affects Chongqing's breathing patterns. The city awakens in a misty haze at dawn as the sun rises over the Jialing and Yangtze Rivers, causing the water to reflect golden hues. The mornings are for the early commuters negotiating the twisting streets, the elderly tai chi practitioners in the parks, and the street sellers erecting their stalls. There is a certain poetic quality to this calm energy—Chongqing before dawn is reflective, a city that pauses for a moment before the bustle starts.

The streets come to life as the morning mist dissipates. The city's monorails wind between skyscrapers and mountains, cutting through the sky. The city's heartbeat is defined by a sensory orchestra made up of the buzz of traffic, pedestrian conversation, and the rhythmic chopping from noodle businesses. However, Chongqing really changes at dusk. The fluorescent heartbeat of the metropolis flickers to life as the sun sets. Like a contemporary impressionist painting, the skyline gleams as it is reflected in the rivers. Illuminated, the well-known Hongya Cave—a multi-tiered architectural wonder carved into a cliffside—attracts tourists eager to take in its unique fusion of modern energy and traditional beauty. Streets turn into rivers of people, moving through shopping malls, night markets, and food stands where the distinct smell of spicy hotpot permeates the air.

Night in Chongqing is simply electrifying. Both locals and tourists congregate at rooftop bars with city views, taking in the breathtaking view of bridges illuminated in a variety of colours. Couples walk down the riverbanks, street entertainers charm bystanders, and tourist boats glide over the water, their reflections blending with the city lights. The waterfront is bustling with activity. The distinction between work and play is dissolved into a fluid flow of movement and energy in this city that defies sleep.

Chongqing is a metropolis that never really stops, whether it is at morning, when the city is occupied by the elderly and silent workers, or at twilight, when it bursts into neon radiance. Time itself seems to have a different beat in this city, which is both meditative and joyous, entrenched in tradition but pulsing with modernity, as revealed by its day-to-night shifts.

15.1.2 Night Markets, Skyline Glows, and the Rhythm of Life

Chongqing at night is a sensory overload in the best possible way. The city's night markets, glistening skyline, and constant bustle create an environment where modernity and tradition meet spectacularly. The street food culture of Chongqing is among the most distinctive features of its nightlife. The aroma of sizzling hotpot, sweet rice cakes, and grilling skewers fills the air at the well-known Nanbin Road night market. In order to attract clients, vendors advertise their specialties and promise the tastiest, spicier bites. Locals and tourists congregate on tables that overflow onto the sidewalks, connecting over freshly roasted chestnuts or bowls of mala tang. These night markets are social hotspots where the city's pulse is at its most unadulterated and genuine, and they provide more than simply food.

The skyline of Chongqing gleams beyond the markets with a brightness that is almost dramatic. The city's expansive, mountainous landscape is outlined by a stunning network of lights created by the illuminated bridges that span the Yangtze and Jialing Rivers. LED displays atop buildings reflect in the water below, transforming the rivers into neon liquid mirrors. Skyscrapers that resemble futuristic sails, such as the Raffles City complex, are emblematic of Chongqing's quick rise to international urban importance. However, the city's nightlife is still firmly anchored in history despite its contemporary sparkle. Stories, Sichuan opera performances, and reunions with old acquaintances over hot tea are all still held in teahouses that have endured the test of time. High-rise bars and electronic dance clubs, on the other hand, serve the younger demographic, demonstrating that Chongqing's nights are open to all, from the avant-garde to the nostalgic.

The way that movement determines life's rhythm is arguably the most intriguing aspect of Chongqing's nightlife. Monorails zip between mountains, taxis sprint over winding highways, and ferries glide along the rivers. The city's location atop precipitous cliffs, separated by sea, guarantees that no one remains motionless for very long. There is a constant undercurrent of life, a sensation of motion, even in the most serene nooks. Chongqing is a city that thrives on constant activity, where tradition and progress, the quiet and the frenetic, the past and the future, all come together to create a symphony of energy. And the cycle starts over when the sun rises.

15.2 Layers of Time: Tradition, Modernity, and the Chongqing Spirit

Chongqing is not just a city—it is a layered experience, where history, geography, and identity intertwine. As this book has demonstrated through urban mappings, Chongqing's multi-layered, multi-identity, and multi-locality character defines its urban fabric.

First, Chongqing is a city of vertical strata [2,9]. It defies traditional urban planning because it was built on steep terrain. Here, historic stilt dwellings sit above contemporary glass towers, and ancient passageways run beneath tall skyscrapers. The intricacy of the city's infrastructure and architecture demonstrates its capacity to adapt to change while holding onto historical artifacts. Second, multi-identity urbanism [10] is embodied in Chongqing. From ancient Ba-Shu civilizations to wartime refugees and modern migrants, it has always been a site of cross-cultural exchange. Because of the imprints each group has left, the city is a living repository of change. The blending of dialects, cuisines, and lifestyles reinforces the idea that Chongqing is not one city but many identities coexisting. Lastly, the city's daily rhythms are defined by its multi-locality. Chongqing is a city of leisure, a tech-driven metropolis, an industrial behemoth, and a historic trading hub all at once. The city reinvents itself through a variety of spatial purposes, whether in the industrial corridors of Liangjiang, the busy commercial zones of Jiefangbei, or the historic neighbourhood of Ciqikou [7].

This book shows how Chongqing functions beyond of predetermined definitions by following these layers over time and place. Tradition and modernity, permanence and transition, remembrance and ambition, all coexist in this city that is always negotiating its past and future [4].

15.2.1 Echoes of the Past in a City of the Future

Chongqing is a paradox: a city that races toward the future while carrying deep echoes of the past. As one of the most historically significant cities in China, it has a deep conversation between history and modernity in its landscape. The historic tea houses of Ciqikou, where calligraphy-adorned walls and wooden beams still reverberate with tales of merchants and academics, are a glimpse of the city's past. Time seems to slow down here. In addition to playing mahjong and sipping tea, elderly people share stories of Chongqing's period as China's wartime capital. The stone-paved lanes, which are lined with stores that date back a century, provide a striking contrast to the neighbouring contemporary skyscrapers.

However, Chongqing's future picks up speed just outside of these old areas. Glass towers and LED-lit skyscrapers with state-of-the-art ads adorn the Jiefangbei key business district. Liangjiang New Area, a smart city zone powered by automation, big data, and artificial intelligence, is one location where the city's ambition is most apparent. Here, with its advanced research facilities and startup incubators,

Chongqing establishes itself as a pioneer in China's digital economy. The development of Chongqing's infrastructure also demonstrates the contrast between the old and the new. Once essential for cross-river travel, the famous Yangtze River Cableway is today a popular tourist destination that represents the city's embrace of heritage while updating its transit systems. In the meanwhile, the city's monorails, some of which pass between buildings, offer a future take on transportation that competently/ Skillfully combines innovation with the difficulties presented by the steep terrain.

This conflict between the past and the future is what makes Chongqing unique, not a contradiction. Chongqing layers its histories, allowing tradition and modernity to coexist alongside one another, in contrast to other cities that discard their past to create room for growth. Even as AI-driven skyscrapers, digital economies, and green urbanism set the route for the city's next era, the echoes of historic marketplaces, wartime shelters, and revolutionary movements can still be heard there.

15.2.2 The People Who Shape the City's Pulse

Beyond its historical and physical components, Chongqing's character is formed by its citizens, who are resilient, enterprising, and closely linked to the development of the city. The inhabitants of Chongqing, from street vendors to high-tech innovators, are the epitome of adaptability and drive. The foundation of the local economy in Chongqing is made up of small company owners and street sellers. Generations of families carry on long-standing culinary traditions in the busy night markets and roadside restaurants, where they serve delicacies like grilled skewers, spicy noodles, and hotpot. These merchants preserve the local culinary legacy in a rapidly modernizing environment; they are more than just businesspeople.

A distinct class of Chongqing's labour, meanwhile, is shaping the city's future in high-rise offices. The city is becoming a center for robotics, blockchain, and artificial intelligence thanks to IT entrepreneurs and pioneers of the digital economy. Chongqing's increasing significance in global technology and smart urbanization is shown in the migration of young professionals to innovation districts such as Yubei and Liangjiang. However, the endurance of the people of Chongqing is what really makes them who they are. The city's history—marked by wartime bombings, forced migrations, and economic reforms—has cultivated a population accustomed to rebuilding and reinvention. Many residents who once lived in traditional hillside homes have transitioned to high-rise living, adapting to the vertical nature of the city.

Both returning residents and migratory workers exhibit this spirit of reinvention. Many people from the countryside moved to Chongqing as it became more urbanized, adding to the city's labour force. Simultaneously, educated Chongqing natives who once left for opportunities in Beijing or Shanghai are now returning, drawn by the city's rising status as a major urban center. Ultimately, Chongqing's pulse beats strongest in its people. Whether preserving traditions or forging new industries, they shape the city's energy, ensuring that it remains dynamic, diverse, and forward-moving.

15.2.3 Debating Chesh Theory*: The Redistribution of Urban and Rural Wealth*

Chongqing is an ideal case study for *Chesh Theory*, which I would like to explain here for the very first time. This theory examines the redistribution of urban and rural wealth as cities expand, which is particularly tangible in developing or fast-growing contexts. Unlike traditional urbanization, where rural residents migrate to cities in search of economic opportunity, *Chesh Theory* suggests a sort of reverse dynamic, which is also very different to inverse urbanization or counter-urbanization effects or concepts. This is a socio-economic shift that happens when cities expand and rural areas are absorbed into the urban fabric. This process can be broken down into a few key dynamics (Fig. 15.1):

1. ***Urbanization and the appreciation of rural property:*** As cities grow, formerly rural areas become urbanized, and their property prices soar. Suddenly, rural landowners who may have been smallholders or low-income farmers are in control of valuable real estate. They become extremely wealthy as a result of selling their land to the government or developers for urban projects. This may differ context to context, where land and property ownership follow certain policies.
2. ***Newly affluent rural inhabitants relocating to the city:*** With their newfound riches, former rural inhabitants frequently purchase apartments or commercial real estate in the city, therefore assimilating into the urban middle or upper class.
3. ***Urban residents selling and relocating to the periphery:*** Long-term urban dwellers may sell their expensive inner-city residences due to rising real estate

Fig. 15.1 Summary of the newly-proposed *Chesh Theory*, as an urbanization phenomenon, and its suggested process in four phases

costs and traffic. Seeking a more spacious and tranquil lifestyle, many use the money to buy land or larger homes (like villas) in the suburban or rural outskirts.

4. ***Traditional migration patterns are reversed as a result of shifting social dynamics.*** Former rural residents move into the urban core as property owners, while long-time urban residents choose suburban or rural lifestyles, resulting in an economic redistribution rather than the usual rural-to-urban migration motivated by economic necessity.

Land acquisitions in Chinese cities like Chongqing, Shanghai, and Guangzhou have made erstwhile rural inhabitants millionaires, demonstrating the phenomena of growing urban areas generating wealth for previously rural citizens/residents. In a similar vein, middle-class urban dwellers are looking for larger homes in the suburbs, while rural communities in Bangkok and Jakarta have made money by selling land as these cities grow. Rapid urbanization in Istanbul, Türkiye, has led to the migration of older city families to suburban compounds and the emergence of a new class of affluent landowners from rural areas.

As a sort of urbanization phenomenon, this theory suggests that rural landowners gain wealth through urban expansion, while long-time urban residents relocate to the periphery. This phenomenon is most noticeable in Chongqing's peri-urban areas, like as Jiangjin, Banan, and Bishan. Land values in once-rural areas have skyrocketed as the city has grown. After selling their homes to developers, many rural families who formerly farmed these areas became overnight millionaires (if not billionaires). They have been able to relocate to the city center, or sometimes inner city areas or communities, and invest in commercial real estate or buy upscale flats/apartments thanks to this infusion of riches. Conversely, longtime city dwellers have sold their core residences and relocated outside due to growing property costs and traffic. Hence, an outward migration occurs through this process. In the outskirts of Chongqing, upscale suburban communities have sprung up, drawing middle-class families looking for more room.

This shift calls into question socio-economic power dynamics and conventional narratives of urban migration. Does landownership determine how wealth is distributed in a new urban development model suggested by *chess theory*? Or does it run the risk of generating new disparities between populations with and without land? I thought sharing this idea in a form of urbanization theory or phenomenon could help us debate this further in future research. This is particularly evident across various parts of Chongqing, especially with its recent expansion toward the northern areas (including, but not limited to, shared or cooperation demonstration zones, etc.). Notably, new discussions have started towards potential agreements to incorporate additional land into Chongqing's administrative divisions (such as Guang'an), extending the city's reach beyond its existing borders. By presenting itself as a metropolis where urbanization and rural prosperity converge in surprising ways, Chongqing offers a practical testing ground for these discussions.

15.3 The City of Water and Bridges: A Dance Between Nature and Urbanism

The natural features of the rivers, streams, and canals that flow through and surrounding Chongqing—often known to as the "City of Water"—are delicately intertwined throughout the city [8]. The city's advantageous position at the meeting point of the Jialing and Yangtze Rivers has influenced its urban growth as well as its historical and cultural evolution. A key component of Chongqing's identity, water presents both opportunities and challenges, and the city's bridges represent its changing connection with the natural world.

With their longstanding connections to the area's transportation and trade systems, Chongqing's waterways are more than just geographical features; they actively contribute to the development of the city. The Jialing River converges in the center of Chongqing, while the Yangtze River, the third-longest river in the world, runs through the city like a lifeline. With the city growing vertically up the slopes and over the riverbanks, these rivers have influenced not only the city's economic development but also its urban form.

Perhaps more than in any other city, Chongqing's bridges are a quintessential representation of the harmony between urbanism and environment. These bridges, which span the city's rivers and numerous ravines, link the different districts and contribute to Chongqing's dynamic contrasts. The city's bridges, which range from classic suspension bridges to contemporary architectural marvels, represent the incorporation of nature into urban development, overcoming the difficult topography of the city while emphasizing its relationship to the natural environment. Chongqing's capacity to blend its natural environment with contemporary infrastructure is demonstrated by the dance between water, bridges, and urban expansion. It displays the adaptability and tenacity of a city that has prospered while preserving a strong regard for the environment.

In addition, Chongqing offers multiple sense of diversity such as language, cuisine, and social interactions. The language, food, and social interactions of Chongqing, a city created by centuries of migration, eloquently reflect its rich diversity. These components serve as indicators of how urban fusion continues to shape the city's residents' changing identities in addition to reflecting the city's rich historical layers. Once a thriving commercial center along the Yangtze and Jialing rivers, the city is now a thriving metropolis with a constantly changing cultural landscape due to the blending of ancient customs and contemporary influences. This section examines the ways in which Chongqing's social spaces, gastronomic offerings, and linguistic diversity all contribute to the city's rich and varied identity.

15.3.1 Language: A Living History of Migration

In addition to being a means of communication, Chongqing's language is a reflection of the city's history of migration and the blending of diverse regional and ethnic influences. The local dialect, Chongqinghua, has a unique tone that distinguishes it from other dialects in the Sichuan region, even though Mandarin is the official language [3]. As individuals from many regions of China settled in Chongqinghua, the language changed, resulting in a hybrid of several linguistic components. The city's tolerance to regional variances and its special status as a crossroads for many communities are reflected in the extensive usage of tonal variants and colloquial idioms.

Over time, a cosmopolitan linguistic environment has been facilitated by the migration of people from neighbouring provinces, especially during periods of economic expansion or conflict. Numerous hybridized idioms have emerged as a result of this linguistic blending, including Mandarin, Sichuan dialects, Cantonese, and the native Chongqinghua. People frequently transition between languages and dialects in casual talks, resulting in a fluid and inclusive communication style that represents the city's residents' adaptability and inclusivity. But the difficulties in preserving a unified municipal character are also brought to light by this linguistic diversity. There are generational gaps in Chongqing since elderly citizens frequently speak their regional languages, whilst younger generations are more likely to speak Mandarin fluently. Language mixing fosters a feeling of cohesion while also subtly highlighting the conflicts between tradition and modernity in a city that is changing quickly.

15.3.2 Cuisine: A Taste of Chongqing's Cultural Melting Pot

Without addressing Chongqing's food, which reflects the city's varied heritage, no study of the city would be complete. Chongqing cuisine, which is well-known for its strong, fiery flavours, has come to represent the city's lively and energetic culture. Hotpot, the city's specialty, is a prime illustration of how food can be used as a platform for cultural expression and social engagement. Due to the extensive usage of Sichuan peppercorns and chili peppers, Chongqing hotpot is known for its strong spiciness, which reflects the city's character: brazen, fiery, and unrepentant.

Because of its status as a transportation and trading hub, the city's culinary culture is a blend of regional and ethnic flavours. A vast variety of regional cuisines may be found in Chongqing today, which has absorbed culinary influences from all around China throughout the years. The city's food scene is vast and varied, ranging from Chongqing-style noodles to Sichuan-style *mapo tofu*. There is a sensory assault of sights, sounds, and smells at local markets and restaurants, which range from street vendors to fine dining establishments. For instance, visitors may sample everything from fried rice and spicy stir-fried foods to dumplings and skewers in the busy Jiefangbei district, which is located in the center of the city and is a foodie's heaven. International fusion food is becoming more and more popular, even though Chongqing cuisine is known for its intense flavours and unusual use of spices. The younger generation has embraced new culinary trends, fusing traditional local foods with Western fast food, as the city modernizes and grows more globally linked. Another indication of the continuous cultural change in Chongqing is this blending of tastes and styles.

15.3.3 Social Interactions: Where Cultures Meet and Blend

Centuries of trade, migration, and urbanization have established Chongqing's cosmopolitan identity, which is reflected in social interactions. You can see a merging of the ancient and the new in areas like Shapingba, which is home to the city's universities and migrant populations. Interactions between residents and visitors, as well as the impact of other cultures, are evident in day-to-day life on the streets. Public parks, local markets, and communal areas act as meeting spots for individuals from various origins, fostering cross-cultural interactions.

In Chongqing, the public space is a vibrant hub of social interactions where individuals from various social backgrounds, ethnic groupings, and geographical locations congregate. These commonplace interactions, whether they take the shape of a friendly conversation at a tea stall, a contentious argument in a university lecture hall, or a gathering of friends over a meal at a street-side hotpot, show how various influences are woven throughout the city. A glimpse of this merging of cultures may be seen at the city's marketplaces, including those in Ciqikou's old town, where local merchants and foreign visitors interact and exchange goods, experiences, and tales.

In Chongqing, cultural fusion extends beyond language and cuisine. It also applies to social customs. While calligraphy and tea drinking are still traditional practices in the area, more and more people from many backgrounds are sharing and celebrating them. By providing a stage for both domestic and foreign artists to exhibit their work, the city's cultural institutions—such as art galleries and performance spaces—play a vital part in promoting social cohesion via the arts.

As highlighted throughout the book, Chongqing's identity is defined by its constant evolution, a dynamic fusion of language, cuisine, and social interactions that reflects the city's history of migration, trade, and urbanization. These aspects of diversity provide a distinctive view into how the people of Chongqing are embracing new influences while also maintaining their traditions as the city grows. The city is a real, breathing illustration of how a wide range of opinions, preferences, and customs may influence urban areas to produce a dynamic and constantly shifting cultural landscape.

15.4 Seasons of Chongqing: A City that Thrives Year-Round

Because of its unusual location at the meeting point of the Yangtze and Jialing Rivers in southwest China, Chongqing experiences unique seasonal rhythms that influence the city's personality all year long. Beyond merely marking the passing of time, Chongqing's seasons—from the foggy winters to the scorching summers and the monsoon rains that pour down the streets—define the city's ambiance, impact daily life, and shape its urban identity. To understand Chongqing, where nature and urbanism live in a delicate, always changing equilibrium, one must understand how the city prospers throughout the year.

15.4.1 Misty Winters: A Time for Reflection

In Chongqing, winter is a peaceful, reflective time of year when a layer of mist frequently clings to the city's hills and rivers. Although not as severe as in northern China, the low temperatures cause an atmospheric shift that affects the country's physical and cultural landscape. A constant in Chongqing's identity, the rivers take on a mystical appearance during the winter months when fog rolls across the water's surface, casting an ethereal veil over the city. The mist gives the city's landscape a sense of mystery and softens the sharp edges of the skyline, making it seem ageless and full of tales.

The streets of Chongqing are quieter in the winter as locals huddle up against the cold. During this season, the city's well-known hotpot culture peaks as residents seek solace in the coziness of communal dining. Winter in Chongqing is a time for closer ties since the colder months also bring about a change in social relations, with interior areas becoming more lively as people congregate around steaming bowls of hotpot. Even if there aren't as many outdoor activities in the city, this time of year encourages reflection and a sense of community.

15.4.2 Fiery Summers: The Heartbeat of the City

In sharp contrast to the tranquillity of winter, summer in Chongqing is distinguished by severe heat and humidity, producing an environment that is both hard and stimulating. High temperatures, high humidity, and high air pollution are the city's "three highs", which make the summer months both miserable and exciting. With daily highs regularly rising beyond 40 °C, the heat frequently approaches oppressive degrees, making outdoor activities challenging during the hottest parts of the day.

But summer is when Chongqing really comes to life, even with the severe weather. The city is alive with activity as both locals and tourists congregate in air-conditioned retail malls or take cover in the shaded alleyways of marketplaces. The noises of street sellers selling refreshing drinks like fruit shakes and iced tea fill the sidewalks, and there is a great demand for eateries that serve hot food. With its spicy Sichuan peppercorns and chili peppers, Chongqing's hotpot achieves a temperature that reflects the summer heat of the city. During the summer, the city's nightlife also thrives, with outdoor events and riverfront bars providing respite from the heat of the day and fostering a lively social scene.

This scorching season is further compounded by the summer monsoon season. The residents embrace the torrential rains because they provide a brief break from the relentless heat. However, because the city's drainage systems are tested and streets flood occasionally, the monsoons also create a new beat in the streets. Residents and the urban infrastructure must be resilient and adaptable to the unpredictable climate that is created by the mix of heat and monsoon rains. In Chongqing, summer is a time of resiliency, vitality, and camaraderie as locals figure out how to not just survive but flourish in the face of adversity.

15.4.3 Monsoons: A New Rhythm on the Streets

In Chongqing, the monsoon season signifies the change from the scorching summer months to the milder autumnal ones. The cityscape is instantly changed by the sudden, frequently torrential rains. Because the streets are slick with water and people move around to accommodate the regular downpours, these rains influence the rhythms of Chongqing's everyday life. During the monsoon season, the city's architecture—which includes a complex system of bridges and steep hillsides—becomes a spectacle as rivers surge and fog rolls in, producing breathtaking views of the surrounding area.

The monsoons are seen as an essential force that restores the city's rivers and adds to the verdant surroundings for which Chongqing is renowned, despite the difficulty they may cause. The city feels rejuvenated by the monsoon rains because the air is cleaner and the parks and streets are cleaned. Because of the monsoons' rhythmic character, the city is calmed, life moves more slowly, and people are more connected to nature. The monsoon season signifies a change in the routine of daily life for the residents of Chongqing. Residents of the city must modify their daily schedules at this time, but they also have time to think as the intense heat has temporarily subsided. The water-slick streets now appear to reflect the rivers, establishing a smooth transition between the natural world and the city. The city breathes in a new beat during the monsoon season, influenced by the interaction of people, land, and water.

15.4.4 A Year-Round City

With its character changing with the seasons, Chongqing is a year-round thriving metropolis. While the scorching summers produce an atmosphere full of resiliency and social vibrancy, the misty winter months bring reflection and a sense of tranquillity. A new rhythm is brought about by the monsoon season's intense rains, which affects both the physical surroundings and how people engage with the city. Chongqing's identity is shaped differently by each season, which adds to the city's dynamic urban landscape, which is as adaptable and flexible as the rivers that flow through it. The city adapts to the changing seasons and thrives through each one while preserving a strong bond with the natural world.

15.5 The Future of the Dragon: Where Chongqing is Heading

Chongqing, a city that has long been characterized by its winding rivers, steep hills, and busy streets, is currently at a pivotal point in its development. Through its city walks, this vibrant metropolis' journey reveals not only the obvious changes but also the significant cultural, social, and environmental transformations that are taking place. As Chongqing develops, it becomes evident that the city's future depends on adopting sustainability and smart urbanism as two cornerstones while preserving its rich cultural legacy.

Our investigation of Chongqing through its city walks reveals a number of important themes, one of which is the city's strong ties to its past and future. It is crucial that the city maintain its connections to the customs and historical sites that shape its character as it develops into a cutting-edge, international hub of innovation and trade. Whether in the lively streets of Ciqikou, the busy marketplaces of Nan'an District, or the city center's historic buildings, the intersections of the old and the new are evident everywhere. These intersections—both literal and symbolic—draw attention to the fine balance Chongqing will need to preserve going forward.

The adoption of smart urbanization by Chongqing will be among its most notable features in the future. The intricate topography of the city, which is characterized by hills, rivers, and fast urban growth, creates both opportunities and problems for the adoption of cutting-edge technology that can improve sustainability, efficiency, and connectivity. Rapid infrastructure development, including the growth of public transit systems, the incorporation of digital services, and the shift to smart mobility, is creating a new urban environment that is more accessible and linked than in the past. Chongqing must use these technologies to build a more ecological, efficient, and liveable urban environment that serves its expanding people as well as tourists if it hopes to succeed as a smart city [1]. The city's well-known bridges, which span its various districts, represent the possibilities for physical and digital linkages in the future and provide a preview of how the city may connect its varied inhabitants and places in ways that are both more inclusive and more efficient.

Even if smart urbanism is crucial for Chongqing's future, sustainability must be incorporated as a guiding principle into the city's growth plan. Concerns over waste management, water pollution, and air quality have all increased as a result of

Chongqing's explosive growth. It is imperative that the city concentrate on sustainable development strategies that lessen its environmental impact while preserving its economic momentum as it grows. The development of Chongqing into a city that coexists peacefully with the environment will mostly depend on the incorporation of green areas, sustainable energy sources, and environmentally friendly infrastructure. Its future will be greatly influenced by how its rivers and waterways are managed, in particular. In addition to serving as the city's physical borders, the Yangtze and Jialing Rivers are essential to its identity, economy, and sustainability. Chongqing can make sure that it is resilient to urbanization and climate change by concentrating on flood control and river conservation.

However, the cultural diversity that makes Chongqing unique must not be sacrificed in the name of the city's continued urbanization [5]. In order to prevent the city's development from erasing its history, Chongqing's distinctive legacy—from its ancient neighborhoods to its varied culinary traditions—must be preserved. Chongqing's essence is found in its local customs, festivals, and cultural landmarks, all of which must be preserved and celebrated amidst contemporary progress. The city's past provides a strong basis for the future, particularly its function as a commercial and cultural hub in southwest China. The concept of the "City Walk" is symbolic of this process—the act of walking through the city allows us to engage with both its old and new dimensions, offering a tactile connection to the cultural landscape that is essential for shaping the future.

Furthermore, the growth of the city needs to be seen from an inclusive perspective. The advantages of Chongqing's economic expansion and growing worldwide impact must be distributed to all facets of society. It will take deliberate policies and actions that support equity to close the gap between the surrounding rural areas and the urban core, which is still marked by major discrepancies. Urbanization must prioritize the people who live in this metropolis as well as the development of infrastructure and technology. The ability of Chongqing to build an inclusive city where all citizens can take advantage of the opportunities brought about by the city's expansion—whether through work, housing, or cultural engagement—will determine how successful the city is in the future.

In conclusion, Chongqing's capacity to successfully negotiate the challenging nexus between tradition and modernity will determine its destiny. The city must be dedicated to maintaining its cultural identity and creating a feeling of community for all of its residents if it is to develop into a smart, sustainable, and globally significant metropolis. As Chongqing keeps expanding, the city walk represents the city's continuous journey, one that respects its history, welcomes the future, and honors the lively, diverse residents. Like its rivers, Chongqing's future will be dynamic, fluid, and always changing, but it will also be anchored in the rich cultural currents that have shaped the city for centuries.

References

1. Cao N, Wen Y, Wang J, Gao X, Sun P (2024) Identifying shrinking cities from a physical city perspective and influencing factors: a case study of the Chengdu-Chongqing economic circle in China. Appl Geogr 168:103314
2. Chen KL, Wu HN, Cheng WC, Zhang Z, Chen J (2017) Geological characteristics of strata in Chongqing, China, and mitigation of the environmental impacts of tunneling-induced geo-hazards. Environ Earth Sci 76:1–16
3. Cheshmehzangi A, Tang T (2022) Chengdu-Chongqing city cluster: From an economic zone to a major economic circle in the Western China. In: China's city cluster development in the race to carbon neutrality. Springer Nature Singapore, Singapore, pp 77–92
4. Cheshmehzangi A, Tang T (2024a) Navigating China's urban revolution: from tradition to transformation. In: China under construction. Springer, Singapore, pp 1–10
5. Cheshmehzangi A, Tang T (2024b) Chongqing: navigating rapid urbanization and regional development challenges. In China under construction: shaping cities through recent urban transformations. Springer, Singapore, pp 81–106
6. Kresse K, van der Krabben E (2022) Rapid urbanization, land pooling policies & the concentration of wealth. Land Use Policy 116:106050
7. Li Y, Cheshmehzangi A, He BJ (2023, November) A framework of urban renewal for leisure space in neighborhood streets: a case study of Shazheng street in Chongqing, China. In: International conference on urban climate, sustainability and urban design. Springer Nature Singapore, Singapore, pp 693–705
8. Li Y, Wei D, Chen Y (2014) The development and evolution of bridge in Chongqing China. Comput Modell New Technol 18(7):166–174, Available online from: http://cmnt.lv/upload-files/ns_39art26_CMNT1807-16_BIN_Li.pdf
9. Xi Y, Li X, Zhu H, Zhang W, Zhao S, Xu W (2022) Three-dimensional high-precision assessment of mountainous urban underground space resources: a case study in Chongqing, China. Tunnell Undergr Space Technol 123:104439
10. Zhu C (2020) A bigger picture: the making of artists in contemporary China. Doctoral dissertation, The Chinese University of Hong Kong, Hong Kong

The manufacturer's authorised representative in the EU is Springer Nature Customer Service Centre GmbH, Europaplatz 3, 69115 Heidelberg, Germany. If you have any concerns regarding our products, please contact ProductSafety@springernature.com

Printed and bound by CPI Group (UK) Ltd, Croydon, CR0 4YY
01/10/2026
02225935-0002